Advance Praise for

For the Life of the World

"Today we are witnessing a global crisis in the relationship between humanity and the natural environment. To face this ecological challenge, what is required is a significant cultural response that draws upon all our resources in a creative and imaginative way. The Christian theological and spiritual tradition has from its very beginnings 'lived' within its landscape, crafting an intimate spirituality in the desert, among the forests, and towards vast horizons of the seas. In *For the Life of the World*, Robin Gibbons makes a fine dialogical contribution, based upon his long familiarity and engaged commitment to the ecclesial, liturgical, and theological traditions of the Christian East. In the polyphony of concerned voices, *For the Life of the World* offers new and ancient perspectives for guiding us along the path."

—Anthony O'Mahony, Fellow at Blackfriars, University of Oxford, and the Sir Daniel and Countess Bernardine Murphy Donohue Chair in Eastern Catholic Theology at the Pontifical Oriental Institute (2018)

"With *Laudato Si'*, Pope Francis placed environmental theology firmly within Catholic social teaching, not only as duty and virtue but because of the ordering of creation and each human's place in society on the Earth out of which we were made. Robin Gibbons shows that this is not just an extrapolation from theology applied to the politics and scientific anxieties of the early twenty-first century but integral to the tradition of both East and West, Catholic and Orthodox. With light from the East as a Melkite Greek Catholic priest, he sets the Church at the forefront of stewardship 'for the life of the world', presenting the spiritual case for its salvation and preservation and thus its future growth towards perfection rather than loss."

—Father Mark Woodruff, Chairman of Society of St John Chrysostom

"*For the Life of the World* by Robin Gibbons is an invaluable book that provides a unique exploration on the urgency for an ecological care for our planet. This book cannot be endorsed enough for its thoroughness and overall challenge. Among the points Gibbons explores is the urgent need for all churches to collaborate together on this issue. This book offers an overall explanation that refers to the rich theological and liturgical tradition of the Eastern Orthodox churches and Melkite Greek Catholic Church since ecological approaches are all inclusive in their worship. Among some of the things explored and explained are the construction of Church buildings with their frescoes and icons, as well as the approach of the iconographer in using the elements of creation.... These are just some aspects of this book that Gibbons thoroughly investigates, explores, and explains. All are a means to show the importance of the churches to collaborate together to increase an awareness of the drastic need to care for our planet instead of using and abusing its resources for our own ends. It is an urgent call to collaborate with the creator in restoring peace, harmony, and unity within ourselves and all creation and at the same time increase that unity in Christ among churches and the whole of humankind."

—Sister Esther Pollak, OSB, Nun at Turvey Abbey and Founder and President of the British Association of Iconographers

For the Life of the World

Washington College

Studies in Religion, Politics, and Culture

Joseph Prud'homme
General Editor

Vol. 12

The Washington College Studies in Religion, Politics, and Culture series
is part of the Peter Lang Humanities list.
Every volume is peer reviewed and meets
the highest quality standards for content and production.

PETER LANG
New York • Bern • Berlin
Brussels • Vienna • Oxford • Warsaw

Robin Gibbons

For the Life of the World

An Eastern Christian Approach to Nature and Environmental Care

PETER LANG
New York • Bern • Berlin
Brussels • Vienna • Oxford • Warsaw

Library of Congress Cataloging-in-Publication Data

Names: Gibbons, Robin, author.
Title: For the life of the world: an Eastern Christian approach to
Nature and environmental care / Robin Gibbons.
Description: New York: Peter Lang, 2019.
Series: Washington College studies in religion,
politics, and culture; vol. 12 | ISSN 2151-7010
Includes bibliographical references and index.
Identifiers: LCCN 2018031967 | ISBN 978-1-4331-5386-0 (hardback: alk. paper)
ISBN 978-1-4331-5387-7 (ebook pdf) | ISBN 978-1-4331-5388-4 (epub)
ISBN 978-1-4331-5389-1 (mobi)
Subjects: LCSH: Climate changes. | Eastern churches—Doctrines.
Classification: LCC QC981.8.G56 G494 2019 | DDC 261.8/8—dc23
LC record available at https://lccn.loc.gov/2018031967
DOI 10.3726/b14543

Bibliographic information published by **Die Deutsche Nationalbibliothek**.
Die Deutsche Nationalbibliothek lists this publication in the "Deutsche
Nationalbibliografie"; detailed bibliographic data are available
on the Internet at http://dnb.d-nb.de/.

29 Broadway, 18th floor, New York, NY 10006
www.peterlang.com

For my Mother, Elizabeth Marie-Jeanne Gibbons (1929–2015), who taught me my faith and love for all living things, and in homage to the on-going work of the "Congregation for Oriental Churches and the Pontifical Oriental Institute (1917–2017)."

TABLE OF CONTENTS

ACKNOWLEDGEMENTS

I would like to thank the following people who have helped me during the writing and publication of this book: The Melkite community in the UK, in whose service I minister and whose spirituality and history have contributed so much to my knowledge throughout my writing. Professor Anthony O'Mahony for his support and wisdom. The Religious of the Assumption, in particular the Kensington community who have put up with me for so long but as my sisters have been invaluable to me in their encouragement and commitment to environmental care. For the witness and prayers of the Clarisses in the Monastery of St Clare in Poligny (Jura, France) and for all my family in that same region of natural beauty who have helped shape my love and concern for the life of the world. For my brothers and sisters, even those who disagree with me, because their discourse has helped me form my own ideas and walk my own path. For Professor Joseph Prud'homme, who kept me to the task of writing this work, and last, but not least, Luke McCord of Peter Lang, whose patience, advice, and presence across the ether has been a sustaining help at all times. I could not have completed my work without you all.

INTRODUCTION

Too often the Christian Church has had an ambivalent approach to our planet, its nature, living creatures and resources. On the one hand there is a strong tradition of stewardship which comes from the Judaeo-Christian scriptures and a command to take care of what resources we have, on the other the implications of being fruitful, multiplying and utilising these resources has not always been a positive experience either for humans or any other living creature, let alone Mother Earth herself! In part this has something to do with a theological approach that sees human beings at the top of the pyramid of life under a Divine presence to which all life owes its origin and continuation. Human beings are intelligent, and so have adapted, created, and existed on this planet using all the materials to hand and making their mark in history. Christianity in both East and West has not always served nature and living creatures well, possibly more in omission than commission. Our scriptures and texts do not immediately urge us to respect animal life for instance, in fact sacrificial activity in the OT suggests more of a substitutional attitude where animals represent humans and take their place. In some cases such as the poor old serpent, there has been a decidedly negative influence! The NT itself does not explicitly state any great ethic concerning environmental issues or animal welfare in terms that we would understand today, later writings are more concerned with the theological context of persons, Christ, Trinity, God and Salvation (to give a few examples) which places the human condition right at the centre of things, and yet looking again at the 'deep' tradition of the Church often opens up old doorways of perception where forgotten thoughts come alive!

In the Christian theology of the Incarnation, the human being is taken up into that divine realm and relationship through God-in-Jesus Christ's enfleshment of the material. In this sense, the '*kenosis*' of Jesus represented in Pauls hymn in Philippians 2: 5–11, that emptying of the divine one to take on human flesh and life is our paradigm of humility towards all things and becomes in our own lives the matrix of serving others, not only an 'emptying out' of self but a 'filling up' of life in God, through his resurrection. In this way the vocation of the Christian is to be part of a conduit, enabling the divine presence to be experienced and perceived in the material world. The Eastern Church has always understood incarnation and resurrection to be the way in which human life transforms and transfigures itself and 'the life of the world'. If the human Jesus is the 'living icon' of God, by extension, humanity, male and female are understood to be icons of Christ. This is further extended by the outpouring, in a direct and unequivocal way, of the Holy Spirit in the Pentecost experience, which is then continued in the life of the Church.

The rich theological tradition of the Eastern Church has always retained that close link with the experiential dimension of faith expressed in liturgy. Scattered throughout the liturgical texts are great examples of a more holistic approach to nature, one has only to visit the Initiation Rites to see the clear connection of symbolic life and nature, water and oil are particularly stressed within this context and the long blessing prayers over the baptismal water connect salvation, the OT Genesis narrative of flood and destruction with the covenant of restoration and life explicitly related to the living action of the life giving Spirit. Initiation is in this tradition not only death to an old form of life, but forgiveness of sin in order that we may conform more to the Divine Life and see in this new life a new creation present at all times and everywhere. In the Eucharist the '*epiclesis*' is a direct outpouring of the Spirit on material elements that become through faith the very life of Christ himself as food and drink for others, a theological understanding of immense power, and dare one say of extreme importance for our attitudes towards all things on this planet and beyond! Again in the liturgical acts of praise and prayer frequent reference is made to the beauty and importance of all created things, our use of psalms and hymns can take us into a very rich and fertile field of discovery where unexpected treasure can be found, reminding us that the divine view of stewardship is much more about careful concern and acknowledgement of individual value, a viewpoint that sees us as good caretakers rather than total users and abusers.

In the regular hours of public prayer we discover how in these acts of common worship the sanctification of the day takes place, and in intercession the needs of the world and all on it are brought before God and ourselves. This is both a cyclical and linear perspective where the daily round of time is connected at particular moments through the assembled community gathered for prayer, particularly the two hinges of the day at Morning and Evening Prayer. We can see this in part of the concluding prayer of the Coptic tradition for any hour of prayer:

> Have mercy on us, O God, and have mercy on us, who, at all times and in every hour, in heaven and on earth, is worshipped and glorified, Christ our God, the good, the long suffering, the abundant in mercy, and the great in compassion, who loves the righteous and has mercy on the sinners of whom I am chief; who does not wish the death of the sinner but rather that he returns and lives, who calls all to salvation for the promise of the blessings to come.[1]

This is very much the direction most churches take, the consecration and offering of each part of the day, it gives a broader sense of concern and care for the whole world and the whole of creation not just humankind. but in the night offices such as Vigils, the focus is linear in that it looks to the *parousia*, the second coming of Christ at the end of time, this too has its theme of everything being brought together in the final act of the new creation where all will be all! The words from the Coptic Midnight Office stress this quasi-joyful attentiveness, waiting for the Bridegroom who is of course Christ. 'O my soul, be mindful of that awesome day, and wake up and light your lamp with the oil of joy, for you do not know when the voice will call upon you saying: Behold, the Bridegroom is coming'. Christian Liturgy is based on the reality of the Risen Christ, called in the phrase of Jean Corbon (a fellow Melkite theologian), '*liturgy de source*' (the wellspring of worship), he wrote:

> Because the Risen Jesus is humanity glorified, he is present through his spirit to every place and age not only as savior, but as saving, not only as Lord, but as priest and sacrifice and victim. Nothing in his being or action is ever past except the historical mode of its manifestation. As the Byzantine Liturgy prays to and of him, 'You are the offerer and the offered, the recipient and gift'.[2]

However there is a connection with the environmental issues affecting the present day, one of the prayers at the Midnight Office of the Syrian Orthodox Church, the *Eqbo* Prayer stresses this:

> O Lord, by the prayers of the chosen twelve apostles, bless the twelve months of the year. May the earth receive regular rain and sun that we may be blessed by the heavenly grace and the harvest of the earth. O Lord, by the rain and drizzle of Your blessing, may the fruits grow and the needy and the poor eat those and praise Your name. *Moryo rahem `a lay noo `adarayn*

So until the Lord comes, the prayer asks for blessing on the seasonal rhythm and for prosperity and plenty for all. This waiting for the Lord, so deeply engrained in the faith of the Christian Church, is not a passive act, for already in Christ creation has been made new, but not yet perfected, so until He comes we, the faithful, must be active in the work of blessing and sustaining, taking on the responsibility of the creator. This might all hark back to a more agrarian world, but nevertheless we cannot dismiss such theology easily. Without it the human person can easily become a despotic religious figure, one 'in charge' rather than 'being in service of'. It is to be like Christ in every way we can. Patriarch Bartholomew wrote this about the disagreements between Orthodox East and the Latins (Roman-Catholics):

> The essential question is this: On the road you follow…have you met Christ and how have you recognized Him? If the answer is yes, then describe Him to me and point out the way. If the answer is no, then tell me why. For my part I would speak to you of my own experience, so that we can help one another as we walk on this road, one leading the other toward Christ, God and man.[3]

In the writings of the eastern theologians there are also surprising references to a much broader and integrated perception of blessing and holiness within creatures and the natural world. For example St Basil has a number of references to all kinds of wildlife in his sermons, speaking of the creation of Fowl and Water Animals he a says when we read the story of creation in Genesis:

> the great wonder of the wisdom of the Creator appears. What a difference He has foreseen among winged creatures! How He has divided them by kinds! How He has characterized each one of them by distinct qualities! But the day will not suffice me to recount the wonders of the air. Earth is calling me to describe wild beasts, reptiles and cattle, ready to show us in her turn sights rivaling those of plants, fish, and birds. 'Let the earth bring forth the living soul' of domestic animals, of wild beasts, and of reptiles after their kind. What have you to say, you who do not believe in the change that Paul promises you in the resurrection, when you see so many metamorphoses among creatures of the air?[4]

There is a striking ease with which he uses these images of creation to draw parallels with the resurrection, metamorphoses they may be, but their use shows a consideration for creatures that instinctively places them in direct location to the Holy One, in Basil's eyes these are wonders of creation, something to be noticed and marveled at, a touch of God's glory. The fact that they also occur in a homily drives home the connectivity. This is not surprising when one thinks of the interior space of the eastern churches, filled at this point in time with icons, mosaics and frescoes, part of which read like the '*biblia pauperum*' of the middle ages, earth lower down, heaven above but all united in the cosmic space of the liturgy and all pointing to the redemptive power of the Most High, who in Jesus calls us to holiness through the life—giving Spirit. Basil literally sings the praise of creation and in significant words tells us that we and they are united in God:

> 'Beasts bear witness to the faith. Hast thou confidence in the Lord?' Thou shalt walk upon the asp and the basilisk and thou shalt trample under feet the lion and the dragon. 'With faith thou hast the power to walk upon serpents and scorpions. Do you not see that the viper which attached itself to the hand of Paul, whilst he gathered sticks, did not injure him, because it found the saint full of faith? If you have not faith, do not fear beasts so much as your faithlessness, which renders you susceptible of all corruption'. It's a pity we haven't heard contemporary preachers pointing out that creatures bear witness to the faith in this manner.

Turning to the ascetic lives of the Desert Fathers and Mothers we find other avenue of exploration, though there is no direct teaching on animals one can discern, particularly in some of their lives, a close identification with animals and birds as a sign of closeness to God, a return to the paradise image of all things connected and in harmony with each other. Strange things can happen in the 'deep' history of this part of the tradition, the Monastic and Eastern stress on fasting is not only a means of self denial, but in refraining from the consumption of animal products there is both a real reminder of paradise lost, where the consumption of animal flesh did not take place, but also of the coming and future Kingdom of God where we will have no need of this kind of food and will be restored into a right relationship with the animal kingdom. The sixth century monk, John Moschos in his travel book, The Spiritual Meadow, describes *dendrites*, monks who live in the branches of trees emulating the birds, *boscoi*, grazers who crawl on all fours eating herbs and grass like the monk Pardos. Monks like Abba Gerasimos who tamed a lion and lived with it or again those who too reduce any semblance of pride became what we call 'fools for God' a feature of this way of life that became much venerated in Russian Orthodoxy.[5]

There are of course examples from the saints who have much to teach us about attitude and ethic, for example, St Seraphim of Sarov is perhaps the Orthodox equivalent of the West's St Francis, his relationship with the wild animals was witnessed by many. At one period in his life he entered deep solitude and retreated into the forest to a hermitage there. Here he befriended the wild animals so that rabbits, lizards, lynxes, foxes and wolves would come to him to be fed. Matrona Pleshcheyeva of the Diveyevo Monastery and Fr Joseph of his own community of Sarov witnessed him feeding a tame bear by hand. It is said of Isaac the Syrian that his humility and likeness to Christ was such that animals were tamed as they approached him, the effects of sin and disunity restored in forgiveness. This is a two fold theme in the East: firstly that the natural world is a way to know God and see salvation through other living creatures so that we become one with the created order, but secondly the 'new creation' in Christ also means harmony with the animals, a harmony that they bring to us as part of the new creation! In this sense we can be reminded of Christ's own example in the desert as told in Mark's gospel. Though many commentaries suggest that reference to the wild beasts heightens the danger of his situation, we could turn that around and see in this comment another way of perception, that the wild beasts recognised the presence of the one who knows and understands them and are in harmony with him. 'He was with the wild beasts, and the angels attended him!' (Mk 1: 13) There are many others whose voices can speak to us of the need to not only repent of our *greedy and rapacious ways before it is too late, but also of the divine presence permeating all* things. It can come as a surprise to find these ancient spiritual teachers reminding us about the necessity of care for other living beings and that they too have a defined place in the creation of God. I wonder how many preachers today would use the example of the humble earthworm as a basis for their homily, but St Basil did to great effect! Perhaps most importantly they speak of the divine love for all things that are created. If we, say the Saints of the East, are to become like God, then we too must have that great quality of mercy, compassion and love for all creation! It is hoped that some of the ancient insights of the Christian East will help us revitalise and rediscover that quality of mercy for our beleaguered world!

Notes

1. Conclusion of Vespers. The Book of Hours, 11th Hour (sunset) Vespers. www.agpeya.org/vespers.
2. Jean Corbon. 1998. *The Wellspring of Worship*. Paulist Press, New York.
3. O'Carroll, M. 1998. *A Light from the East*. Queenship Publications, Santa Barbara, CA, p 114.
4. Basil. The Hexameron. Homily VIII. 8.The Creation of Fowl and Water Animals.
5. See Wortley, J. (trans). 2012. *The Spiritual Meadow*. Lysimakos. pp 69–86.

· 1 ·

TOWARDS A SPIRITUALITY OF ENVIRONMENT

I. The Need for a New Spirituality of the Environment

There is no doubt that paradigmatic shifts are taking place in the Earth's climactic structures. We hear much about the effects of Global Warming, even though there are many who still disregard the signs, but it is now difficult, too many reports of issues connected with human misuse of resources emerges each week. The Natural Resources Defence Council of the United States, which was founded in 1970 to protect land and water from human pollution and greed, constantly issues reports into the various problems facing our planet and its inhabitants.[1] This is only one of many global groups now convened to tackle these issues urgently, but also actively seeking to provide solutions that benefit all parties. The NRDC has over 1.4 million members and can call on over 500 professionals, including lawyers and scientists. There are many others! A recent article in various newspapers drew attention to the plastic garbage in the ocean that is finding its way into the marine food chain and causing many problems, in particular cases the garbage comes together to form a kind of plastic soup. Perhaps it is because 'out of sight' does mean 'out of mind' that we fail to recognise the precariousness of so much we take for granted. John Vidal in an article for The Guardian newspaper in February 2015 drew attention to the call for the UK to create the world's largest conservation area for marine life in either the Atlantic or Pacific region. Why? Because the UK, through its various territorial dependencies, has charge of the 5th largest area of ocean in the world, 6.8 million square kilometres a fact many in the UK will not have known! At present the political will does not have much environmental credibility, but voices are getting significantly louder and our moral responsibility for the future is huge.[2]

Even in a local situation such as the area of Great Britain in which I live, West Oxfordshire on the edge of the Cotswolds, there is enough evidence from the changing patterns of the weather, more rain, less harsh winters, very hot summers, within which there are shorter patterns of extreme weather. In the area of wildlife the decline in certain species of 'native' birds, as well as in numbers of the seasonal migrations of birds such as Swift and House Martin; and in small mammals such as the Hedgehog, as well as the almost extinction of our native Adder in the region, act as warnings of something bigger taking place. There are of course people who will invariably challenge the conception that 'Global Warming' is in fact happening at all, and point to the patterns of the past and recurrent cycles of weather over periods of time as a question mark over the whole issue. Despite increasing evidence to the contrary and the projected problems over water and food with a growing global population, the verdict is still out as far as they are concerned. And yet the evidence continues to mount inexorably!

What is particularly worrying is the fact that some religious leaders, often speaking from a personal viewpoint, have caused some confusion about the Christian response and attitude towards this problem. As far back as February 18th 2007, in an article in the Australian Sydney Sunday Telegraph, the then Roman Catholic Archbishop of Sydney, criticised those who talked up global warming and suggested that this was something that had been vastly exaggerated. In his own blunt style he wrote that it was a 'nonsense' that 'aimed to frighten' and was the product of 'Global warming doomsdayers'. He returned to this theme in another article in the same paper on the January 6th 2008. Here he portrayed himself as a sceptic and sought to criticise those who portrayed the worst case scenario as 'scaremongers', later making some particularly trenchant comments about the environment being for some, a 'cause that is almost a substitute for religion'.

The Archbishop did however acknowledge that it was the duty of Christians to be concerned about the environment and was prepared to accept that the whole situation and the scientific data being produced was complicated. Though it was fortunate that he was writing and speaking in a private capacity, and therefore his remarks do not carry any ecclesial weight, they are nevertheless the public utterances of a public person and the common perception is that behind them lies official Church teaching. In fact a number of Church People have called his remarks irresponsible and at variance with his own Church's increasing stance on this issue. In the light of further developments and with the approach of Pope Francis who has made climate change issues one of the thrusts of his ministry as Bishop of Rome, this stance seems deeply problematic and ironic, although it is symptomatic of a number of Catholic Bishops (and other Christian leaders) world wide. In fact the Pope's Encyclical *Laudato Si* expressly sets the Catholic Church and others on a path which cannot ignore the issues facing us. He invites all peoples to consider their 'home':

> More than fifty years ago, with the world teetering on the brink of nuclear crisis, Pope Saint John XXIII wrote an Encyclical which not only rejected war but offered a proposal for peace. He addressed his message Pacem in Terris to the entire 'Catholic world' and indeed 'to all men and women of good will'. Now, faced as we are with global environmental deterioration, I wish to address every person living on this planet. In my Apostolic Exhortation *Evangelii Gaudium*, I wrote to all the members of the Church with the aim of encouraging ongoing missionary renewal. In this Encyclical, I would like to enter into dialogue with all people about our common home.[3]

Nevertheless whilst the Archbishop's negative opinions are still an all too familiar one, and these provocative ideas continue to dominate headline space, other faith people, and this does include religious leaders, are quietly at work (and often unreported) encouraging people to concern themselves responsibly with environmental issues. Particularly in the 'western world' where a cult of individualism and consumerism holds sway, the voice of anybody who purports to hold different values, especially spiritual ones, is important. Christianity, not least because of its long intellectual and spiritual tradition, could perhaps and should be in the forefront of any engagement, even though the Christian community has also to acknowledge their part in the destruction of life over the centuries. It was noticeable that the South African Council of Churches in 2009 issued a Report on Climate Change in which they emphasised the difficulties and challenges facing the churches including the problem of finding a moral voice on this issue:

> The 'prophets' who are typically issuing warnings about climate change do not do so in the name of Christianity or even from a religious perspective. Those who have taken the lead and have called for moral vision and moral leadership include scientists, consultants, politicians and journalists. Their work is in fact being undermined by (religious) prophets of doom and destruction who typically evoke fear, not hope. This leads to an inability to confront the stark challenges ahead. Moreover, the driving forces behind economic globalisation are often associated with a vocal form of right-wing Christianity. Christians in South Africa and elsewhere in the world therefore find themselves in an uncomfortable position. They are being addressed in a prophetic mode instead of exercising their own prophetic responsibility. Moreover, as many churches have had to admit, climate change seldom receives a priority on the social agenda of the church. This implies that Christians cannot speak about climate change with any degree of moral authority.[4]

There is the difficulty, this agenda is not often the primary focus for church mission and some leaders feel it must take second place to other social concerns. However it is becoming abundantly clear that climate change and environmental issues are indissolubly bound up with many other problems including justice. The Catholic journal The Tablet published an article by Francis McDonagh on February 25th 2015 entitled, 'When the dream becomes a nightmare'. This illustrated the catastrophic impact that an unrestricted and destructive extractive culture has done to Brasil's ecology, climate and economic growth. Brasil was linked with China, India and Russia as one of the great emerging economic powerhouses, now its economy is crippled by environmental issues amongst other things, mainly concerned with water, drought, pollution and flood caused by deforestation amongst other resource problems:

> '2015–16 is likely to be catastrophic,' warned a group of scientists, who wrote an open letter to politicians during last year's election campaign but were ignored. The leader of the group, Professor José Galizia Tundisi, even invoked Winston Churchill. Brazilian political leaders, said Tundisi, should follow the example of Churchill in 1940 'and tell the population brutally that the crisis is very serious'.[5]

Brasil is not alone, there are too many numerous examples across the globe that openly connect human impact on environment and climate with other political and economic problems.

II. Hope From the East

Despite this mess there are those, particularly amongst Christians, who continually exercise their prophetic mission to keep the issue before our eyes, the Ecumenical Patriarch Bartholomew is one, he has made it part of his outreach to draw attention and awareness of ecological and environmental issues. At the 2014 Interfaith Summit on Climate Change in New York he said:

> unless we change the way we live; we cannot hope to avoid ecological damage. This means that—instead of solely depending on governments and experts for answers—each of us must become accountable for our slightest gesture and act in order to reverse the path that we are on, which will of course also include prevailing upon governments and leaders for the creation and application of collective policy and practice.[6]

It is with a sense of urgency that Christianity needs to address the issue of our environment seriously. It is imperative that it should become a major focus, not only of theological endeavour, but of local church and community concern. Whilst there are encouraging signs, there are underlying problems, one of which is a particular need to help shift one focus of our spiritual tradition to encompass a 'spirituality' of the environment. The reason for this is to bring to another level of human experience a different theological and spiritual awareness and also a holistic perception of our place on this planet, not only as the human family but as part of creation itself. One that is not only concerned with hard evidence, scientific data and statistics, but also with that deeper understanding of the heart and spirit, in order that a deeper 'conversion' of the human family might be achieved.

One particular historical 'tradition' of Christianity might be able to help us, that of the Oriental and Eastern tradition, understood here in the Eastern Catholic, Orthodox and Oriental experience and sense of faith and belief that is lived, accepted, loved and passed on anew in each generation especially in gatherings for prayer and worship. This ancient tradition fundamentally recognises that history is not incidental to, but is the context for, God's work with creation. In one sense this is nothing particular to these churches, because it is incumbent on all Christians to share in the responsibility for care, concern and consideration for our planet, but the Oriental and Eastern Christian has particular insights and experiences to share, which are qualitatively different from those who come from the Latin and Protestant communities. Again this is not problematical, the Christian faith is not opposed to history, and does not see any opposition between religious and world history, but it does recognise the cultural shifts and patterns of different ages and the need for a contextual response. Hence in the past, 'spiritual traditions' have arisen from particular situations in response to social, economic, political and religious needs. We also acknowledge that within the eastern part of Christianity there has always been an particular understanding of 'tradition' being the living transmitter of collective wisdom, what in the Latin church we can call the '*sensus fidelium*', a slightly opaque but engaged theological concept in which the life of the church plays an important part in formation of faith and living theology, the context often being that of prayer and worship where the adage '*lex orandi, lex credendi*' remains true, meaning that the relationship between the praxis of worship and life, and the theology of the church still holds firm![7]

Belonging to an historical milieu does not deny the enduring legacy of the historical continuum, which is something the Eastern Churches readily understand, as many of these communities, especially in the Middle East, come from the very inception of Christianity: after all at the heart of the Christian faith is the primary historical-faith event of the resurrection of Jesus Christ, an encounter which is continually being experienced anew in every generation of church members. However in examining the spiritual legacy of other ages we cannot be totally uncritical, the question of interpretation and adaptation remains, for what was necessarily good in the life and times of a fourth century ascetic in the Egyptian desert may not be so in the quite the same way for a twenty first century city dweller in Rio de Janeiro. The various levels of symbols, the '*verbum visibile*', as well as textual language of such traditions will need to be carefully decoded in the context of each particular cultural setting without losing the essence of its original meaning. This means a complicated and critical dialogue with spiritual symbol and text left by these traditions.

As Philip Sheldrake puts it so well: 'Despite the wariness of many people in the contemporary West about institutional religion, the place of history in the study of spirituality is a reminder of the positive power of the religious-spiritual "tradition"'.[8] It is a deeply held conviction of mine, that in this twenty first century, a renewed spirituality of space and place, built onto and partly taken from the wider traditions, will be a means of transforming peoples values and life styles to enhance, rather than destroy our environment.[9] It may also act as a uniting force helping the fragmented Christian Church draw nearer together.

III. Green Prophets From the East

That there is a shift in perception amongst Christian church leaders becomes more evident when we consider the growing number of those who are working to draw people's attention to the plight of our planet and want to encourage positive action. Chief amongst them, as already noted, is the Orthodox Ecumenical Patriarch Bartholomew 1 of Constantinople, often termed the 'green' Patriarch. He has been one of the unsung heroes of the new movement, active in voicing concern about global warming and continually pointing out the human responsibility for destruction of habitat, species and climate by our tendency to innate consumerism and greed. In 1997 the Los Angeles Times (November 9th 1997) carried an article on this aspect of his mission with the headline, 'Harming Environment is Sinful, Prelate Says' (November 9th 1997). He has continually worked to focus attention on a need for radical conversion, in June 2003 for instance, he called together a gathering of scientists, political leaders and journalists, to join him on a cruise ship to the Baltic to see first hand the results of over fishing and to discuss marine preservation and the hazards of over fishing amongst other concerns.

It is not surprising that this should be one of his great missions. Patriarch Bartholomew comes from the Eastern Orthodox tradition in Christianity with its strongly monastic and liturgical spirituality. At the heart of this is a deeply rooted sense that through the Incarnation, God in Christ, has entered into the life of the world making it holy and through his Resurrection has transformed it, enabling us to reach God and in our own baptismal mission, gifted by the Holy Spirit, reach out and help transform our world. Christ has linked together the life of the world with the life of God, but it also means that the each human being must change their way of life to allow this. In the Orthodox and Oriental spiritual tradition their theology gives to any understanding of human place in created life a deep sense of the importance of all and of creation as something that is part of the life and love of God, as Metropolitan Anthony Bloom put it;

> The vocation of man, from the first chapter of Genesis, has been to sanctify the whole creation of God. St Gregory Palamas tells us man was created belonging to two worlds; the world of God and the world of matter...God took what is the basic material of creation and made man out of it, so that man participates in everything that was created out of dust, beginning with the smallest atom and ending with the greatest galaxy, and all the rest of the visible created world around us.... This has enormous importance. If god became man in Christ, then Christ participates, as does

> each of us, in the dust, in the galaxies, in the atom, in the animal world. He has taken upon himself the experience of all createdness. He is one of us, but in him everything created can see itself.[10]

This great theological insight is not only discovered in texts and writings, but in the way the Orthodox-Eastern Christian reverences and uses the material world, especially through their worship and prayer, which is understood as God's blessing and gift, something given to us for careful use and consideration, and not for our domination and exploitation. There is the virtue of humility in this approach which places humanity, in terms of connection and interdependence at basic levels of existence, on the same level as all created life, not above it! One thing it does is call us to a sense of vocation, we are called to be less 'stewards', an image taken from biblical account of creation (Gen 1: 28),because unfortunately this is a term open to abuse, and more 'carers'. In some contexts the biblical interpretation of creation in term's of human activity, places the 'steward' as somebody standing over and above those in his/her care. To be a 'carer' is to become linked with others as 'friends' of all creation, an image taken from the command of Jesus, as Johns Gospel puts it:

My command is this: love each other as I have loved you. Greater love has no-one than this: to lay down one's life for one's friends. You are my friends if you do what I command . I no longer call you servants, because servants do not know their master's business. Instead, I have called you friends, for everything that I learned from my father I have made known to you. (Jn 15: 12–15)

That metaphor is, I believe, part of a radical re-appraisal of our vocation in the context of environment and life, the servant image works, but only so far, because as the scriptural Jesus puts it so well; 'So you also, when you have done everything you were told to do, should say,' We are unworthy servants; we have only done our duty (Lk 17: 10). Amongst others 'carer' is certainly a richer and more nuanced word.

Whilst many of us no longer have servants for our household tasks as people did in past ages as a matter of normative experience, all around us the richer people of the world, still utilise people as 'servants' even if they are not described as such. One cannot make value judgements based on generalities and some employers are fair and follow the statutory legal requirements of employment and care, these people are good employers, but there are too many other stories of abuse and deprivation where the servant is in reality a 'slave', in some cases becoming a species of stateless person, whose identity, because their legal documents are confiscated and held by their employers in order to prevent them leaving, makes them a 'stateless person', placing them in an unfair and illegal dependency situation, where the wages earned are either minimal or used up on necessities for survival. There are frequent reports in the media and International Organisations such as Amnesty International continually draw attention to the vulnerable status of such workers.[11]

What this suggests is that we need some new models of service for the Christian, particularly the servant image, which has its limitations and has overtones of exploitation. For me the particular problem is that a servant has to do what they are required and that is duty, as Jesus suggests, but if we examine the friend and carer model, we can see that something changes in the arrangement of persons, for not only does friendship/caring go further in creating dialogue and understanding, it also implies an equality especially of empathy with people, this is not only good for us but it also creates a responsible environment for growth in mutual understanding and a sympathetic approach to the other in an evolving relationship. If we transpose these images into the context of creation and the environment we have to acknowledge that our friendship/caring with life has not always been positive. In the context of the environment and living creatures we, the human species, have too long treated them as our 'servants', who are there for our bidding, but time is running out as more and more of these species disappear, for more than ever it is essential and urgent that the faith communities reiterate and proclaim God's love for all life in a new approach to ecological and environmental issues, and one that respects the mutuality of creation with ourselves. Too often we recognise the importance of something after it has gone!

In his Message for the day of the protection of the natural environment on September 1st 2006, Patriarch Bartholomew wrote these words which encapsulate part of this spiritual awareness; 'The ultimate cause of all this behaviour is man's (*sic*) egocentrism, an expression of his self-willed alienation from God and his effort to make himself god.'[12] Here then is a starting point we can all recognise, believer or not, and it is the response we make to this need for radical conversion which can form the basis of a spirituality for the environment, eco-spirituality, call it what we will. The way forward as the Patriarch writes in the same letter is to raise awareness in order that people may act: 'let us make every effort possible, each from where God has placed us, to rein in our reckless over consumption'.[13]

His ecumenical rapprochement with the Vatican has managed to bridge at least one gap between the Roman Catholic and the Greek Orthodox Churches, which is an historic achievement in itself and perhaps a signal for the future. If spiritual leaders can be advocates of environmental issues, then their traditions can also facilitate new visions for the future in which the spiritual plays as much part as the material. Due to the Patriarch's efforts, Pope John Paul II became involved in collaborating with projects dealing with environmental issues and it has become part of the Vatican agenda. This is not an insignificant development, as the policy of mutual involvement and concern has continued under the Papacies of Benedict XVI and of Pope Francis. On the occasion of a joint service between Patriarch and Pope Benedict XVI at the Patriarchal Cathedral of Saint George in the Phanar, Istanbul, on November 29th 2006, Bartholomew's address to the Pope stressed the need for dialogue and encounters of love and truth between Christians, whilst reminding us all of the need for a common understanding which one can reasonably assume will also concern global issues. This, together with the abuse of religious freedom and open persecution, is perhaps one of the most pressing ecumenical issues of our age!

For the Patriarch, speaking about the need for unity between Roman Catholics and Orthodox, the historical divisions in Christianity are 'equally accountable for the steps—just, of course, as we are for any missteps—along the journey and in our struggle to obey the command of our Lord that his disciples "may be one".'[14] In the past this would not have been an easy thing for traditional Orthodox or Catholics to readily accept; certain conservative factions within both Churches would still reject this kind of statement but if they do, they are turning their backs on the voice and call of the what Pope John XXII called 'the signs of the times', which in the process of spiritual discernment we can identify the call of the Holy Spirit. One of the latest initiatives of Patriarch Bartholomew was to invite religious, scientific and political leaders to a symposium in September 2007. In a well documented event, they went to look at the *Ilulissat* glacier in Greenland, which has now developed moulins, holes 10–15 metres across in the last few years, a definite sign of dramatic change and in view of the glacier they prayed together. He has been a continual activist, if one might use such a term, in drawing attention to the issues of environment, human factors in its destruction and the need for human beings to have a profound change of heart and life. Despite the failure of the 2009 Copenhagen Conference on Climate Change, there is a sense of urgency and palpable unease in the political world. Regions such as the coastal areas of Philippines and Vietnam are amongst some of the most vulnerable places in the world. It is all too evident that something has drastically changed in climatic effects. According to the World Wide Fund for Nature, this Philippine area is considered as one of the most vulnerable to climate change. Impacts range from extreme weather, periodic inundation, droughts, destruction of habitat, homes with resultant scarcity of food. For the Filipino community, climate change has been a present, constant and unpredictable reality, those living in coastal communities and the lower rung urban communities amongst the most vulnerable. These places are not alone across the globe you will find many other communities in much the same state.[15]

Pope Benedict XVI did not write an encyclical on the Environment but he gave a consistent and prophetic voice on environmental issues. He too deserves the epithet alongside Pope Francis as a 'Green Pope' particularly by his concern for the issues of climate change and the resultant profound and destructive impacts on the planet, but more especially because of his concern for the poor and vulnerable, not only humans, but other living creatures. In his World peace Day Letter for January 1st 2010, Benedict had this to say:

> We are all responsible for the protection and care of the environment. This responsibility knows no boundaries. In accordance with the *principle of subsidiarity* it is important for everyone to be committed at his or her proper level, working to overcome the prevalence of particular interests.[16]

That statement is certainly at variance with the point of view of people like Cardinal Müller and Cardinal Raymond Burke amongst others. For whilst the Pope warned against, what he termed, the exaltation and absolutization of nature over human persons, his emphasis on the reciprocity between people and nature places a different onus on integrative responsibility, and in some way points (or perhaps hints) at a deeper tradition of loving stewardship within the Christian faith tradition, one which I would consider is profoundly influenced by Eastern Christianity.

His successor Pope Francis continues this ecumenical partnership in working for the planet with Patriarch Bartholomew, and has placed environmental issues at the forefront of his own ministry. Patriarch Bartholomew continues to address these ecological problems in constant letters, addresses and encyclicals whilst Pope Francis issued an Encyclical on the Environment in 2015. *Laudato Si* commits the Catholic Church of both East and West to an unprecedented witness of what Patriarch Bartholomew calls, 'mutual respect among human beings and all created beings as well as with everything that has life and offers life,'[17] and should become one of the most powerful calls to action, not least because of the media attention which any major statements from the Vatican attracts, but also because, one hopes, if the Pope truly collaborated with other church leaders across all the divisional boundaries as well as reaching across the inter-faith barriers, he would then be articulating a common sense of purpose and vision on behalf of all other religious leaders and peoples.

These global issues are not about fundamentals of doctrine, nor are they expressions of particular church identity; they concern the future health of all peoples and life on this small planet and beyond. *Laudato Si* and any future Papal document on the environment cannot simply be a document for the Catholic Church, even though this Church contains within its structures all known nationalities and many Eastern Churches in communion with Rome. It would have to be truly ecumenical. In this case the Pope would have to dialogue with others such as the Ecumenical Patriarch and other Patriarchs, the Archbishop of Canterbury, other Church leaders and faith leaders such as the Dalai Lama and others of good will, in order to produce something of immense significance for the whole future of the earth, and its inhabitants. The Vatican held a conference on climate change in April 2007, where on the April 27th Pope Benedict XVI stated that we need to not only respect creation, but focus on the needs of sustainable development. The Guardian noted that '(at) a Vatican conference on climate change, Pope Benedict urged bishops, scientists and politicians—including UK environment secretary David Miliband—to "respect creation" while "focusing on the needs of sustainable development".'[18] Together with Patriarch Bartholomew's commitment to the environment, this message from the Pope represented gathering momentum in the Catholic Church both East and West, and indicated movement towards significant changes in approach, but a note of caution needs to be sounded, not everybody agreed, there are still the sceptics and deniers of this world! But there is hope, in *Laudato Si*, Pope Francis acknowledged the role of Patriarch Bartholomew, and by implication the whole theological position of the East. Praising the Patriarch he gives expression to the debt a greedy and unheeding Church, let alone world, owes to the vision of this leader. The particular section of the Encyclical where he talks about Bartholomew's great contribution is worthwhile quoting in full because it highlights many of the perspectives the East brings to this great and continuing discussion:

> I would mention the statements made by the beloved Ecumenical Patriarch Bartholomew with whom we share the hope of full ecclesial communion. Patriarch Bartholomew has spoken in particular of the need for each of us to repent of the ways we have harmed the planet, for 'inasmuch as we all generate small ecological damage', we are called to acknowledge 'our contribution, smaller or greater, to the disfigurement and destruction of creation'. He has repeatedly stated this firmly and persua-

sively, challenging us to acknowledge our sins against creation: 'For human beings... to destroy the biological diversity of God's creation; for human beings to degrade the integrity of the earth by causing changes in its climate, by stripping the earth of its natural forests or destroying its wetlands; for human beings to contaminate the earth's waters, its land, its air, and its life—these are sins'. For 'to commit a crime against the natural world is a sin against ourselves and a sin against God'. At the same time, Bartholomew has drawn attention to the ethical and spiritual roots of environmental problems, which require that we look for solutions not only in technology but in a change of humanity; otherwise we would be dealing merely with symptoms. He asks us to replace consumption with sacrifice, greed with generosity, wastefulness with a spirit of sharing.[19]

This of course comes from his own deep rootedness as a monk formed in the traditions of prayer and reflection. Here is no ascetical unworldly hermit, but as we shall note later when we consider the monastic contribution to environmental issues, here we have somebody deeply rooted not only in concern for the world, but in God's concern for the world. This is because this life is a life lived in totality, as Benedicta Ward put it so well when speaking about the desert monks and nuns it is: 'the hard work and experience of a lifetime of striving to re-direct every aspect of body, mind, and soul to God, and... (T)hat, also, is what they meant by prayer: prayer was not an activity undertaken for a few hours each day, it was a life continually turned towards God.'[20]

IV. Ecological Concern an Imperative of Mission

Whilst it has always been the case that religions have as part of their core values and teaching a deep respect for the earth, practice has often lagged significantly behind when it comes to placing the human being within a larger context of responsibility for its actions on the rest of life! We continue to waste material resources and turn our face away from the destruction and extinction of life. It is heartrending to discover just how fast many of our planet's species are moving towards extinction, a recent list compiled for *USA Today* at the beginning of 2015 gives 15 species that could have been extinct at the end 2015![21] I'll give one simple example to show what happens when we ignore the issue. The International Rhino Foundation has these statistics for the Black Rhino, in 1970 there were 65,000 in Africa, by 1993 the number had decreased to 23,000, in 2013 the catastrophic effects of hunting and poaching had lowered the animal numbers to just 5,005 and in 2015 to 4,848. These magnificent beasts are on the critically endangered list, but worse still is a sister species the White Rhino which in 2015 has only five animals left in the wild! There are so many others, as well as habitat and plant life that we destroy continually.[22] Paul Brockelman puts it succinctly in his work 'Cosmology and Creation' when he points out the innate selfishness and short-sightedness of so much of our human worldview: 'what we have lost then, is the ability to see our lives as part of a wider order and reality beyond our daily and passing self-centered desires and dreams. By seeing nature and the entire universe as a "stuff" put here for our endless productive transformation and use, we have reduced reality to a mere extrinsic value for us; it is no longer encountered as intrinsically valuable *in itself*. As a consequence, we have lost any sense of belonging to a larer and more significant drama and reality'.[23]

In an address to youth gathered at Sao Paulo in May 2007 Pope Benedict XVI condemned the deforestation of the Amazon and in his 2008 letter for World Peace Day urged people to focus on the environmental issues confronting the planet: '... it is essential to "sense" that the earth is "our common home" and, in our stewardship and service to all, to choose the path of dialogue rather than the path of unilateral decisions. Further international agencies may need to be established in order to confront together the stewardship of this "home" of ours; more important, however, is the need for ever greater conviction about the need for responsible cooperation. The problems looming on the horizon are complex and time is short. In order to face this situation effectively, there is a need to act in harmony. One area where there is a particular need to intensify dialogue between nations is that of the *stewardship of the earth's energy resources*'.[24]

The Archbishop of Canterbury Rowan Williams, convened a meeting of Church leaders in October 2009 just before the Copenhagen Summit and issued the first statement of faith leaders in the UL on climate issues:

> We recognise unequivocally that there is a moral imperative to tackle the causes of global warming. This is reinforced by the reality that it is the poor and vulnerable who are most profoundly affected by the environmental impact of climate change—especially drought, floods, water shortages and rise in sea levels. Faith communities have a crucial role to play in pressing for changes in behaviour at every level of society and in every economic sector. We all have a responsibility to learn how to live and develop sustainably in a world of finite resources.[25]

All these events are signals that those in the religious traditions of the world are waking up to an immense responsibility, the task not only of re-educating, but also of educating those who see nothing wrong in continuing to carry on as they have done before, or to develop things without any thought of what this might mean for others, but in all of this the hidden problem seems to me to be an issue with many of the Churches themselves who are often seen as arrogant consumers whose track record needs careful examination. Here I believe the Eastern Church has much more to offer, for these communities, apart from perhaps the Russian Orthodox Church at a certain stage of her life, have not had the same influence in politics and statecraft nor in 'imperial' missionary endeavours over the last few centuries, as have the Roman Catholic and Protestant traditions. For many of them either under the Ottoman or Islamic rule, their historical status as '*dhimmi*', that is second class citizens subservient to others, gives them another voice, and their traditions echo a deeper cry from the heart of creation itself.[26] They are the humble churches in a very real sense, and have lived in poverty and simplicity. One great characteristic of their liturgical spirituality is *koinonia*, that is fellowship, communion with others (1 Cor 10: 16; Phil 6.) This can be seen in the approach and witness of Patriarch Bartholomew himself, as Fr John Chryssavgis wrote: '…what is perhaps most characteristic of the Patriarch's initiatives is the mark of humility. The Ecumenical Patriarch is able to see the larger picture. He recognizes that he is standing before something greater than himself, a world before which he must kneel, a chain that long predates and will long outlast him. Therefore, he speaks of self-emptying (*kenosis*) (Phil. 2.4–11), ministry (*diakonia*) (Luke 10.40; Acts 1.17, 25; 6.4), witness (*martyria*, a term which also has the sense of martyrdom and suffering) (John 1.7, 19), and thanksgiving (or *eucharistia*, a term which also implies liturgy) (Acts 24.3; 2 Cor. 4.15).'[27] These elements of eastern theology and spirituality, together with a deeply rooted monastic presence, which gives a particular quality to the eastern Christian life, all weave together a powerful gift of symbol, praxis and outlook, to help us construct a new hermeneutic of exploration in that sense of *koinonia*, fellowship, enabling us to interpret things in a different way and approach the whole question of our relationship and responsibilities towards the planet and its life in a more open and humble manner. St Anthony of Egypt in his texts on saintly life sums this insight up in these succinct words:

> The world is maintained through God's providence and there is no place where providence does not touch. Providence is the self-fulfilling Word of God, the Giver of form to the substance constituting this world, the Architect and Artist of all that is …. To a man who believes and desires, it is in no way difficult to apprehend God. But if you wish also to see Him, look at the perfect order and providence for all things, which have been and are through His Word.[28]

One doubts if the 'perfect order' of Anthony's inspirational words still remains in the same condition today as it did for him in the fourth century, but these words bring hope, not only of our ability to change, to convert from bad ways, but also still learn to respect and care for the environment even at this eleventh hour!

If the planet is to somehow continue, then people like St Anthony need their teaching about respect for creation to be not only accessed from its place in a particular understanding of human religious history, but be retrieved and made widely available as part of a received and living tradition, which is seen as necessary for the very continuance of our own and other's existence. Whether one believes Religions are the best vehicles for this type of re-education or not, they are nevertheless powerful vehicles for crystallising and communicating expressions and ideas. Christianity, based on the understanding of Jesus as Redeemer, has perhaps a unique role to play in this reappraisal of relationships within the world order. With its highly developed system of thought and its living tradition of prayer and worship, the various strands that have exerted such profound effects in peoples lives can be reinterpreted to help us all focus on a humbler and more compassionate role for humankind on planet earth. Part of this task will be to retrieve the voices and teachings of the past and develop a spirituality of the Environment. In this case, newer theologies and older movements can be brought together, not to create a synthesis but to provide a framework for people's spiritual journeys.

In this respect I suspect that the Eastern Christian tradition, with its deeply liturgical spirituality, its great emphasis on the role and transformative work of the Holy Spirit and its connection to, and love of the material world, will be a big quarry to mine for those who seek to guide others. This prayer, said at the beginning of each liturgical action in the Byzantine tradition, encapsulates the essence of this: 'Heavenly King, Consoler, the Spirit of truth, present in all places and filling all things, the Treasury of blessings and Giver of life, come and dwell in us, cleanse us of all stain and save our souls, O Good One.'[29] It is very much a prayer at the heart of Eastern Christian perception, one that recognises the all enveloping role of the Spirit in all life, a gift that consoles, heals and renews the face of the earth. But there are others, men and women loved in both east and west, whose message ties in with this praise of God in all creation. One of the best loved is of course, Francis of Assisi, that thirteenth century mendicant, founder of the Order that bears his name, but somebody with that quixotic mixture of harshness towards self and gentleness to others. He treated his own body harshly yet recognised the interrelation of all created things, far from being a 'romanticised' vision of the created world, Francis' understanding is more complex and fits in with that spirituality of presence in all places and things revealed in the byzantine prayer.

In Francis, as in the eastern tradition, we discover that a perception of a deeper, mutual and interdependent relationship between all the elements of creation. His famous Canticle on Creation, 'The Canticle of the Sun' (*Il Cantico del Sole*) expresses a belief in the harmony between all the elements, linked to both an understanding that in Christ all are one, all creatures and elements except one, are brothers and sisters, for Francis it is through our 'Mother Earth' that reconciliation becomes the only way to achieve peace. This a translation of the Umbrian text of the original codex, the words forcefully convey Francis' intensity and love of God in all things:

> Most high, all powerful, all good Lord! All praise is yours, all glory, all honor, and all blessing. To you, alone, Most High, do they belong. No mortal lips are worthy to pronounce your name.
>
> Be praised, my Lord, through all your creatures, especially through my lord Brother Sun, who brings the day; and you give light through him. And he is beautiful and radiant in all his splendor! Of you, Most High, he bears the likeness.
>
> Be praised, my Lord, through Sister Moon and the stars; in the heavens you have made them, precious and beautiful.
>
> Be praised, my Lord, through Brothers Wind and Air, and clouds and storms, and all the weather, through which you give your creatures sustenance.
>
> Be praised, My Lord, through Sister Water; she is very useful, and humble, and precious, and pure.
>
> Be praised, my Lord, through Brother Fire, through whom you brighten the night. He is beautiful and cheerful, and powerful and strong.
>
> Be praised, my Lord, through our sister Mother Earth, who feeds us and rules us, and produces various fruits with colored flowers and herbs.
>
> Be praised, my Lord, through those who forgive for love of you; through those who endure sickness and trial. Happy those who endure in peace, for by you, Most High, they will be crowned.
>
> Be praised, my Lord, through our Sister Bodily Death, from whose embrace no living person can escape. Woe to those who die in mortal sin! Happy those she finds doing your most holy will. The second death can do no harm to them.
>
> raise and bless my Lord, and give thanks, and serve him with great humility.[30]

The space and place of the world, revealed as our 'Mother' and our 'common home', is a sacred space in which this cosmic fraternity and sorority have to take place.[31] Generally the Greek Fathers have been happier to use maternal metaphors about Christ in its tradition, this coupled with the veneration of Holy Wisdom, which is feminine in principle, can help shift our focus away from metaphors of patriarchal dominance.[32] One beautiful early Christian hymn composed by Clement of Alexandria (150–215 CE) sings of Christ not only as bride, teacher, king but also as wing, rudder, shepherd and because he is Word of the all-highest father (Clement hints in his use of the word children) sings of the 'maternal' Christ, who is Emmanuel, God-with-us! The fact that this was permitted to be sung at the '*agape*' meal gives it a greater resonance as a hymn of created love:

> Bride of colts untamed
> Wings of unerring birds,
> true rudder of ships,
> Shepherd of the royal lambs,
> collect your
> Simple Children
> So that they extol with holiness,
> Praise guilelessly
> With innocent mouth,
> Christ the teacher of children,
> King of the holy ones,
> All-conquering Word
> Of the highest father,
> Master of wisdom.[33]

The problem is that one species, the human, has not often seen fit to recognise this wisdom of interconnectivity and humility and has treated the other elements as items for appropriation, and has used religious text and tradition to justify this way of life. This is why this type of ancient hymn can really help us recover a more balanced viewpoint. The human centred approach can no longer be a viable way of existence, for at the heart of our tradition is another voice, which we can rediscover and use, and which tells another story. So the question we can ask ourselves constructively is; in what ways can perspectives and insights from our eastern Christian tradition help us create better understandings?

A holistic vision requires at base level, the removal of threat, particularly that of war and disasters of the natural world, often caused by our interference. Societal peace is somehow linked to a scriptural image of ecological utopia. Mark Brett, writing about procreation and the politics of identity in the book of Genesis, reminds us that the first Covenant in Genesis (Gen 9: 8–17) reveals the divine promise encompassing all creatures, 'not only is there no licence for Israelite separatism in Genesis, even the distinctiveness of the human species is undermined when humankind has to share the divine vocation of co-creation with the earth and other creatures'.[34] Brett also points out that the prophet Hosea (Hosea 2: 18) shows that the security of the new covenant with God in that ecological utopia, envisages not only an end to conflict, but also humankind's active covenant with the 'beasts of the field', the 'birds of the heavens' and the 'creatures that move on the earth'.[35] This is a very different vision of human supremacy, one in which that context of friendship becomes the foundation of trust instead of the inflated version of human hegemony. The hidden voices from the east can help us find another identity in our tradition, one which is 'constituted by innumerable traces of other identities'.[36] One place we can find them very much alive, even in the darkest places of a war torn region, is in the context of worship!

Notes

1. See website and report at http://www.nrdc.org/oceans/plastic-ocean/faq.asp.
2. See Vidal, John. *Conservation Call for UK to Create World's Largest Marine Reserve*. The Guardian, February 10th 2015. http://www.theguardian.com/environment/2015/feb/10/conservationists-call-for-uk-to-create-worlds-largest-marine-reserve.
3. Pope Francis. 2015. Encyclical *Laudato Si*, Vatican, Rome, June 19th 2015. n 3.
4. *Climate Change-A Challenge to the Churches of South Africa*. South African Council of Churches Climate Change Committee. Marshaltown, November 2009 SUN Media, Stellenbosch, p 5.
5. McDonagh, Francis. *When the Dream Becomes a Nightmare*. Article in The Tablet, February 28th 2015.
6. Bartholomew, Patriarch, *Message*. *New York Interfaith Summit on Climate Change*. New York, September 21–22nd 2014. Text found at http://www.oikoumene.org/en/press-centre/files/PatriarchmessageInterfaithClimateSummit.pdf.
7. '*the law of prayer (is) the law of belief*', attributed to prosper of Aquitane.
8. Sheldrake, P. 2007. *A Brief History of Spirituality*. Blackwell, London. p 9.
9. See Gibbons, R. 2006. *House of God: House of the People of God*. SPCK, London.
10. Metropolitan Anthony of Sourozh. 2008. *The Living Body of Christ*. Darton Longman and Todd, London. pp 128–129.
11. Articles and reports that reflect the growing evidence for such abuse of power can easily be found, here are two simple reports that illustrate the type of problems hidden; Fisk, Robert. The Independent. Thursday October 12th 1995. *Gulf Maids in Slavery to Reign of Terror*

in Gulf Homes. Ai-Jen Poo. The Guardian, February 18th 2015. *Female domestic workers emerge from the shadows to fight abusive employers*.

12. Message by H.A.H. Ecumenical Patriarch Bartholomew on the Day of Prayer for the Protection of Creation (01/09/2006) prot. No 856. Text found at website of the Ecumenical Patriarchate; https://www.patriarchate.org/-/septon-patriarchikon-menyma-epi-tei-hemerai-tes-proseuches-hyper-tou-physikou-periballontos-01-09-2006-.
13. *Ibid*.
14. Address, Phanar. November 29th 2007.
15. See: Neil Adger, W. *Social Vulnerability to Climate Change and Extremes in Coastal Vietnam*. World Development 27, no. 2 (1999). Pergamon, Elsevier Science Ltd. pp 249–269.
16. Benedict XVI. *Message of His Holiness Benedict XVI for the Celebration of the World Day of Peace. January 1 2010*. Vatican, December 9th 2009.
17. Patriarch Bartholomew. Encyclical Letter for 2015.
18. Vidal, John and Kington, Tom. *Protect God's Creation: Vatican Issues New Green Message for the World's Catholics*. The Guardian, Friday April 27th 2007.
19. *Laudato Si* nn 7,8,9.
20. Ward, B. 1985. *The Sayings of the Desert Fathers: The Alphabetical Collection*. Cistercian Publications, Kalmazoo, Michigan, USA. p xxi.
21. http://www.usatoday.com/story/news/world/2015/01/02/globalpost-species-near-extinction/21182221/.
22. http://www.rhinos.org/rhinos/black-rhino.
23. Brockelman, Paul. 1999. *Cosmology and Creation: The Spiritual Significance of Contemporary Cosmology*. OUP, Oxford. p 11.
24. Benedict XVI. *Message of His Holiness Pope Benedict XVI for the Celebration of the World Day of Peace 1 January 2008. The Human Family*, A *Community of Peace*. Vatican, December 8th 2007.
25. See more at: http://rowanwilliams.archbishopofcanterbury.org/articles.php/770/faith-and-climate-change#Statement.
26. See: Gibbons, Fr Robin. 2014. Essay: *The Dhimmi: Dhimmi and Dhimitude in the Ottoman Empire*. The Living Stones Year Book 2014. *Christianity Engages with Islam: Contexts, Creativity and Tensions*. Meselinde, London. pp 87–106.
27. Chryssavgis, Fr John. *The Green Patriarch: Ecumenical Patriarch Bartholomew and the Protection of the Environment*. 4. Ecumenical Patriarchate Official Website https://www.patriarchate.org/the-green-patriarch. See also Chryssavgis, Fr John. *Ecumenical Patriarch Bartholomew: Insights into an Orthodox Christian Worldview*. The International Journal of Environmental Studies 64, no. 1 (2007), pp 9–18.
28. Kadloubovsky, E. and Palmer, G.E.H. (trans). 1969. *Early Fathers from the Philokalia*. Faber and Faber, London. p 37.
29. Text of prayer taken from *The Divine and Holy Liturgy of Saint John Chrysostom*. Librarie St Paul-Alleluia Press, Jounieh, Beirut. 2004. p 12.
30. Umbrian Text:
    Altissimu onnipotente bon signore. tue so le laude la gloria e l onore et onne benedictione.
    Ad te solo altissimo se konfano. et nullu homo ene dignu te mentouare.

Laudato sie mi signore cum tucte le tue creature spetialmente messor lo frate sole. lo quale iorno et allumini per loi. Et ellu e bellu e radiante cum grande splendore. de te altissimo porta significatione.
Laudato si mi signore per sora luna e le stelle. in celu l ai formate et pretiose et belle.
Laudato si mi signore per frate uento et per aere et nubilo et sereno et onne tempo. per lo quale a le tue creature dai sustentamento.
Laudato si mi signore per sor acqua. la quale e multo utile et humile et pretiosa. et casta.
Laudato si mi signore per per frate focu. per lo quale ennallumini la nocte. ed ello e bello et iucundo et robusto et forte.
Laudato si mi signore per sora nostra matre terra. la quale ne sustenta et gouerna. et produce diuersi fructi con coloriti fiore et herba.
Laudato si mi signore per quelli ke perdonano per lo tue amore. et sostengo infirmitate et tribulatione. beati quelli ke l sosterranno in pace. ka da te altissimo sirano incoronati.
Laudato si mi signore per sora nostra morte corporale. da la quale nulla homo uiuente po skappare. guai a cquelli ke morrano ne le peccata mortali. beati quelli ke trouara ne le tue sanctissime uoluntati ka la morte secunda nol farra male.
Laudate et benedicete mi signore et rengratiate et seruiteli cum grande humilitate.
Pulignani, Michele Faloci (ed.). 1888. *Il Cantico del Sole di San Francesco di Assisi. Foligno:* Tipografia di Pieter Sgariglia. pp 10–11. Translated by Bill Barrett from the Umbrian text of the *Assisi codex*. See http://www2.webster.edu/~barrettb/canticle.htm.

31. *Laudato Si* n 218.
32. See Bynum, C.W. 1982. *Jesus as Mother: Studies in the Spirituality of the High Middle Ages*. University of California Press, Berkeley. p 125 ff.
33. Hengel, M. 1995. *Studies in Early Christology*. T&T Clark, Edinburgh. p 251.
34. Brett, Mark. 2000. *Genesis: Procreation and the Politics of Identity*. Old Testament Readings. Routledge, London. p 27.
35. Brett, p 29.
36. Ibid, p 44.

· 2 ·

LITURGICAL INSIGHTS

I. Worship as a Matrix for Change

In the Eastern worship traditions of the Christian Church there is a rich seam of praise and thanksgiving that places the human in a different form of connection and relationship, not only with God and the world, but also with the cosmos. One has only to look at the iconography of the liturgical environment of the byzantine tradition to see the cosmos and the world of nature replicated in the design and imagery of the frescos, mosaics and icons of the church building itself. This imagery interweaves with powerful correctives in liturgical texts and prayers that prevent those who engage in these acts of worship from becoming too self-centred or too focused on the human condition, for instance, at the prayer of the 'Little Entrance', when the book of the Gospels is solemnly processed through the Church to re-enter the icon screen and enter the 'Holy Place' the priest prays; 'Master and Lord, our God, who have established in heaven the ranks and armies of angels and archangels to minister to your glory: at our entrance, make the holy angels enter also, to minister with us and with us to glorify your goodness'.[1] Immediately the connection between the divine and human is rea' sed, this is perhaps more important than it first seems, because for the B· ntine Christian the Gospel Book is also seen as the vehicle of Christ, the One, who is the Word made flesh encountered in the words proclaimed in His name at the Gospel, a modality or vehicle of Christ's presence in created matter, in this case the book. After the Gospel Book has blessed the entrance the Priest and choir sing the hymn ,

'O son of God who are risen from the dead…'[2] a theme which underlies and underpins this transitional quality of the Divine Liturgy as a matrix for transformation and encounter between the realm of God and human beings.

That does not mean to say there isn't a touch of anthrocentricity in the practise of certain communities, it is all too easy to remain focused on the immediate necessities of life and of contemporary concerns within the doctrinal or moral framework of the Church's teachings. On a personal level I have to ask myself, how often do we hear a string of prayers which indicates great and loving concern for the human condition in all its aspects, local, national and international yet leaves out the darker side of the story for the other non-human living creatures and species of the wider planet? How often do we hear collective and regular prayers of contrition and intercession for the rape and pillage done to the Earth or the mass annihilation of species, a factor and tragedy of contemporary life which can be laid at the feet of each of us. The words of the Byzantine Eucharistic Liturgy call us back to that timeless focus, at the Liturgy of thanksgiving all pray;

'We thank You, Master who love mankind, Benefactor of our souls, for having today made us worthy of your heavenly and immortal mysteries. Make straight our paths, establish us in your fear, guard our life, steady our footsteps, through the prayers and supplications of the glorious Mother of God and ever virgin Mary and of all your saints'. This is a constant reminder that though we are all complicit by omission or commission in the vast destruction taking place, at least we have a voice, but who speaks up continually for other forms of life or the fabric of the world itself? How often have the voices of our preachers been raised in concern for God's other forms of creation? There have been some, but precious few in comparison to those who preach on the concerns and matters that are humanly focussed. Yet in prayers such as the one just quoted we find there is a voice that speaks beyond the years, a constant homiletic of wider horizons, that is the presence of the Divine One acting through the liturgy itself.

II. Transmission of Presence

At the heart of all our words of worship are the texts of the Old and New Testament, especially parts of them that Christians have used on a regular and repetitive basis for their acts of prayer. On a regular basis many eastern communities of Christians are familiar with several sets of scriptural texts that constantly bring world, creature, the need for repentance, and the omniscience of God's rule before us, these are the tradition of the psalms and canticles found in that round of daily common prayer, particularly in the morning and evening celebrations, the particular lessons from the Hebrew Scriptures and the New Testament, read at these daily offices and at sacramental celebrations such as the Eucharist. Also in a particular manner, the words of Jesus proclaimed by ministers such as the Deacon or Priest and heard, perhaps a better expression might be 'received', by the faithful in the constant proclamation of the Gospel; particularly at the Divine Liturgy (Eucharist) which are continually exposing the faithful to the the dynamics and challenge of the 'Good News' of the Kingdom for our lives in the here and now. This approach to the actual physical and audial presence of the Gospel, is one that challenges the community to transform what is heard into an encounter of their faith in Jesus' present amongst them in that realised promise, 'For where two or three gather in my name, there am I with them' (Mt 18: 20).

It also places the transmission of the word and its reception into a clear theology of presence, which is part of the liturgical understanding of the various modalities of Christ's presence (a presence which we also have to acknowledge is part of the charism of the Holy Spirit) found in people such as the ecclesial community gathered for worship, the clergy who minister and administer the sacramental life of Christ for others, the communion of Christ, oil, water, light manifestations of the Spirit in Christ at work in the life of a person.

> The Fathers, notably St. John Chrysostom, applied these words of Jesus (Mt 18: 20) to the liturgical assembly, asserting that it implies the presence of the Lord. The presence of Christ is bound up with the sign (symbol) composed by the meeting of the baptised in one place for prayer.... This presence, however, is not sacramental in the strict sense of the term. The assembly must hear the word of God, which brings about another presence of the Lord: the assembly is orientated towards the sacraments, the act of Christ, and above all toward the Eucharist which makes the glorious humanity of the Crucified really present.[3]

This is a very useful pointer in which the liturgy and its sense of presence can help us to formulate new ways of engagement with ecology and environment. When we start to link the theological insights of the liturgy with our own approach to the cosmos and world about us, we must recognise that these modes of presence can take us forward in three directions. The first is personal presence because the immediate presence of Christ is an encounter of faith between the risen Lord and ourselves, so that our first encounter with this symbolic presence is transformation of our heart ,the *metanoia* required by the Gospel.[4] The prayer after the Litany of Peace sums this approach up succinctly: 'O You who have granted us to pray together in harmony, and who promised that when two or three are gathered to call upon your name, You will give what they ask: do You now fulfil what your servants asks, so far as it is good....'[5]

It is easy to translate that into an insight for the meeting of the twenty first century person with their world, for, whatever we might like to think, deep conversion has to be at the heart of any meaningful connection with the planet, each other and life. We encounter them, but is it an encounter of loving presence? Hardly! That is why the conversion of heart is absolutely essential for any action on our part to do with the environment.

Secondly there is mediated presence, recognition that the presence of Christ is never immediate but depends on that reflective encounter, so that by these symbols we engage a bit more with the deeper issues of commitment and care and are then transformed. In this way we need to reflect more on an ontological approach to creation.

Thirdly it is about the way in which we perceive and encounter the symbolic world in which we live, in other words, not so much a question of seeing our planet, world, creatures as an 'object' to be examined, or simply a presence that is there as a fact, but a presence to encounter, to know and understand. That requires some fairly radical approaches on our part, but it is certainly my contention that the treasury of Eastern Christianity not only has spiritual insights to give, but tried and tested ways to provide pragmatic and practical help in reengaging our attention.

All these moments of conversion, mediation, and engagement are particularly prescient in the eastern tradition, where the insistence on immanence and transcendence of God is so strong so that there is not the hard and fast demarcation of different worlds (so to speak) as may be found elsewhere in Christianity. It might be useful to show this in a simple example, the priest's prayer before the proclamation of the Gospel in the Byzantine liturgy expresses this connectivity between symbol and presence succinctly, but it also allows us to discern the particular theology of transfiguration and transformation. Those three directions are certainly here in the prayer, before we hear and encounter the Gospel, Christ first has to encounter us, almost by direct invitation in that first phrase; 'shine in our hearts', but the dialogue continues and is seen to be rooted in a very deep sense of trust. For 'love' (which is the action of the Holy Spirit) is mediated in our hearts and minds by the 'pure light of divine knowledge' which in turn allows us to then fully engage our hearts and minds and spirit with the Gospel so that in the world we may bring glory, praise, blessing:

> Shine in our hearts, Master who love mankind, the pure light of your divine knowledge and open the eyes of our mind that we may understand the announcing of your Good News; set in us the fear of your blessed commandments, so that, trampling all carnal desires, we may begin to live according to the spirit, both willing and doing everything for your pleasure. For You are the light of our souls and bodies, Christ God, and we render glory to You....[6]

III. God and Creation

These insights help us discover how useful it is to identify areas within these scriptural and liturgical texts which constantly bring before us, in a more reflective and symbolic manner, their deep and abiding concern for all of creation. This is very important because we do need to seek out and retrieve the voice of the God, particularly in that loving dynamic of the Holy Trinity, the Triune God of the Christian tradition, where the creating, redeeming, loving activity of the three persons is so carefully expressed, and then encountered, in the life of the worshipping community. Fr Alexander Schmemann puts this dynamic of the Trinity as community so eloquently:

> 'We believe in one God, but not in a God who is alone, not in a God who is self-centered, not in a God who lives in and for Himself. God is love, Christianity says. But to love is to be turned towards the Other, and at its most exalted, to love is to surrender to Him. The Father, says the Gospel, loves the Son and gives everything to Him. The Son, says the Gospel, loves the Father and gives Himself totally to Him. Finally, the Holy Spirit is this gift of love itself, this love itself as gift, joy, fullness, existence, thirst and quenching of thirst, giving and receiving.
>
> And through this love, through this giving, God creates the world, calls each of us into life, makes each of us beloved and bestows His love, His Spirit, so that each can partake of divine Life. Thus we pray to the Spirit: 'Come and abide in us'.... And

> in answer He comes to abide in us, to lead us, as Christ promised, into all truth…to give us eternal life.
>
> This then is the radiant mystery of Trinity, the God who is Love.[7]

That is very much a summation of Byzantine Theology, reiterating the importance the Trinity places in the spirituality of these oriental and eastern Christian traditions. It also helps us understand their preoccupation with tradition as a means of living out through the ages, the primary encounter in faith as that of the Disciples with Christ in the Resurrection, which in cosmic terms transforms history and therefore creation, but continues in the life of the world through these moments of presence. The Catholic Church (which holds together both eastern and western traditions in a certain ecclesial sense) speaks to all Christians, through its document The Dogmatic Constitution on Divine Revelation of the Second Vatican Council (*Dei verbum*), is particularly succinct in its definition of this on-going activity particularly of living tradition:

> This tradition which comes to us from the apostles develops in the Church with the help of the Holy Spirit. For there is a growth in the understanding of the realities and the words which have been handed down. This happens through the contemplation and study made by believers, who treasure these things in their hearts (cf Lk. 2: 19, 51), through the intimate understanding of spiritual things they experience, and, through the preaching of those who have received through episcopal succession the sure gift of truth. For as the centuries succeed one another, the Church constantly moves toward fullness of divine truth until the words of God reach complete fulfilment in her.[8]

Whilst Christians of various denominations might not agree entirely about the Catholic and Orthodox insistence on the pre-eminence of the role of Bishops in the Church, as teachers par excellence, and fortunately the east gives a timely check and balance balance by paying more attention the work of theologians especially lay ones: nevertheless here is a clear exposition of developmental theology contextualised in the continuity of faith by an ecclesial community. What it gives us is a framework of reference and a sense of s continuum, that like creation itself, the community represented by this ecclesial group, does have a chance to do something constructive for the life of the world and also has a vision rooted not in human prowess and progress but on a totally different focus that of divine truth itself. We keep returning to the ancient and deep roots of a synthesis between all that forms what we call the 'blue planet' as well as the unfolding mysteries of cosmos and space.

IV. Conversion, Creation and Covenant

There in the text of Genesis describing the very activity of creation, we return again to the words of the first account of creation in the Hebrew scriptures to discern (and discover) hints of a manifest presence of God experienced in at least three distinctive ways, God who is the mover of creation by architectonic deign, word that names and forms, but lives only when that breath moves into it. This is of course a particular insight, an interpretation from the East, but a powerful one at that for it engages us in a dynamic approach to creation, it is of course not creation *ex nihilo*, for there with God is the formless void, but one that has attributes of great depth and darkness!

'In the beginning God created the heavens and the earth. Now the earth was formless and empty, darkness was over the surface of the deep, and the Spirit of God was hovering over the waters' (Gen 1: 1).

Suddenly in powerful imagery, the Creator reaches out into the void of nothingness and speaks the Word, that voice and name bring into form the created, 'And God said, "Let there be light", and there was light' (Gen 1: 3) which is then blessed and animated by the living gift of the Holy Spirit. This is made explicit in the second account of creation (Gen 2:4–25) for there the Spirit that hovers over the waters is breathed into the dust as the breath of life to form human kind. But this is not to place humans on a pedestal of priority, there is, if we dig a little deeper, another narrative. The first Covenant in Genesis (Gen 9: 8–17) points to the divine promise encompassing all creatures, not just humanity. 'Division is perhaps hinted at in the idea that the plants and animals were "created according to their kinds" (Gen: 1 vv 11, 12, 21, 24, 25), and a hierarchy certainly arises with the creation of the species that exclusively is made in the image of *Elohim*...(Gen 1: 26–8) ... But even here, where a basic distinctiveness of humankind is asserted, there is an element of continuity between humankind and other creatures. The first divine speech that contains an imperative is addressed to the creatures of the sea and the birds of the air...That is the blessing for procreation has been shared with these creatures'.[9]

The a problem of hierarchy, but that is not insurmountable, this can be addressed in a more constructive way when we consider the divine vocation in Christ and the models he gives us, especially of humility! After all one of his great phrases is that the 'first shall be last and the last first'!

If we examine the image of the group God makes covenant with after the floo', then we also see an inter-connectivity between species, which makes the human equivalent to the animals, for all are part of a covenantal kinship group! If you also look at Genesis 1, so often used to explain human dominance, there is quite clearly no social conflict, all is good, all is blessed, all are to increase and multiply, Human expansion is not at the expense of others! Isaiah's vision of a new heaven and earth removes the remembrance of 'former things' (Is 65: 17) and in that utopian setting of restored harmony the animal and human's treat to each other is removed; 'The wolf and the land will feed together, and the lion will eat straw like the ox, but dust will be the serpent's food,. They will neither harm nor destroy on all my holy mountain' (Is 65: 25). Hosea looks upon the death of living creatures with horror but envisages God ending war and covenanting with all creatures, where peace in society is firmly linked to an ecological paradise. For Hosea ecological destruction comes from violence and war; 'Because of this the land dries up, and all who live in it waste away; the beasts of the field, the birds in the sky and the fish in the sea are swept away' (Hos 4: 3). Restoration comes with peace, in a land where righteousness, justice, love and compassion rule. 'In that day I will make a covenant for them with the beasts of the field, the birds in the sky and the creatures that move along the ground. Bow and sword and battle I will abolish from the land, so that all may lie down in safety' (Hos 2: 18).

This vision is not simply a static text kept within a framework of exegetical study, in the liturgical life of the eastern church these texts are heard as lessons and sung as portions of hymns during the course of the liturgical cycle and are linked to that theological understanding of the New Covenant celebrated in the worshipping community, particularly the Eucharist. In this way they live! This is something deeply rooted in the liturgical theology of the eastern church, Alexander Schmemann describes this as 'epiphany' in which the people become participants in the event they celebrate:

> The Church is not an institution that keeps certain divinely revealed 'doctrines' or 'teachings' about this or that event of the past, but the very *epiphany* of these events themselves. And she can teach about them because, first of all, she knows them; because she is the experience of their reality. Her faith as teaching and theology is rooted in her faith as experience. Her *lex credendi* is revealed in life.[10]

If we are to ever win back a respect and love for the world from the human community, it can only come about by a thorough and ongoing conversion, but as the Scriptures proclaim, this can only take place when the reluctant Prophets of the world go out and recall people back to the source that is God. In the creation accounts of Genesis there is a pattern of worship at work, it isn't simply the description of several aspects of creation or a theology of what that means, it is also an act of worship. At the end of the passages from Genesis once the human beings are finally created, God blesses all that is made, and it is good, 'God saw all that he had made, and it was very good' (Gen 1: 31). We can notice with some amount of contrition, that in these accounts all creation is good, not just the human centred world. The problem comes when, in the Judaeo-Christian tradition (and wider Abrahamic tradition), the apex and pinnacle of creation is directly and inextricably linked with human kind itself. The Genesis story of Noah is one comment on the sinful destructiveness of human beings and what happens when they are left to their own devices, but the destruction of the flood leads to a covenantal relationship with the Creator that renews the promise of love and care, not only with Noah and his family, but all life:

> Then God said to Noah and to his sons with him: Now I establish my covenant with you and with your descendants after you and with every living creature that was with you-the birds, the livestock and all the wild animals, all those that came out of the ark with you: never again will all life be destroyed by the waters of a flood; never again will there be a flood to destroy the earth (Gen 9: 8–11)

This of course is a passage utilised in the context of blessing of water at the baptismal liturgy, particularly at the pouring of the oil, a clear and direct liturgical link to the new covenant that is being made through the re-birth of the catechumen in the waters of the font and their bringing to life in Christ! 'Sovereign Lord and Master, God of our Fathers, Who did send to them in the Ark of Noah a dove bearing a twig of olive in its beak as a sign of reconciliation and salvation from the Flood, and through these things prefigured the Mystery of Grace; and thereby have filled them that were under the Law with the Holy Spirit, and perfected them that are under Grace.'[11]

In some ways it is inevitable that the Christian understanding of the Incarnation should draw the lines of thought towards this view of the new covenant replacing the old or rather being fulfilled in Christ. The battles that took place in the first millennium of Christianity over orthodox Christology are also battles about salvation and the place of the Divine One in earthly life. Nevertheless, the east warns against too much of a direct focus on Christ without the interwoven dynamic of the Trinity especially the Holy Spirit. This has been a particular problem of the west a very 'christo-centric' approach that places the human at the apex of creation. The Eastern Church especially through its liturgical use of the scriptures, recognises the subtle and definite prophetic tradition which forms an insistent and powerful corrective to this imbalance. If we examine the liturgical traditions of our past we can discern strong evidence of a more nuanced approach, one that needs a greater voice in the future. Part of this hermeneutic of restoration will include a stronger approach to the role of the Spirit throughout the pattern of the liturgical use of the Old and New Testaments. This of course brings its own hazards, especially in inter-faith dialogue, nevertheless, if the Eastern tradition's insistence on the role of the Holy Spirit is to be taken seriously, especially in the context of worship, the exploration of the work and role of the Spirit can provide a structure from which to dialogue.

> ...(I)t is the theology of love, implicitly at work, that determines my personal position in the world and also my relationship to others, under the inspiration of the Holy Spirit. Otherwise, I shall end up with a system that has no foothold in the real. Moreover, the aims of the renewal set afoot in the theology of the liturgy are to display man as an active member of the Church, as the dynamic carrier of the living Tradition. The miracle of Pentecost should be a sustaining force at all times, and a constant call to the Church, so that each one may spread the news of God's marvels and tell of the might of the Spirit, as the needs of the times demand....The great ascetic, Nilus of Sinai, in order to demonstrate the sacred character which should accrue to the believers relationships in society, reminds us that we should see the image of the Creator in others, whether they are men or women. 'Your neighbour should be another god after God'.[12]

The Spirit gives life to all and is the animator of life as well as the gift-giver, but this includes all of the cosmos and in earth bound terms all of life. That there is a distinction in the *charismata* the Spirit gives is evident in the way we as human beings perceive ourselves, Paul's great list in 1 Corinthians includes much that one cannot say with any certainty exists outside the human condition, but, and here I posit a great but, there is a difference in examining those kind of statements and explaining them as belonging to humanity and stating that no gifts exist in other life forms, when there is the potentiality for forms of *charismata* to exist. Let me explore this a little more in relation to the Genesis text. If as the author of Genesis puts it, humans have mastery and dominion over all, and animals and life are given to humans so we may live (in a definitively human existence) what then of the beasts that see us as submissive, part of their food chain and so readily devour us, why no mention of that type of phenomenon? Perhaps this is because the brutish and cruel belong to that sense of fallen and sinful people.

There is a corrective, it shows itself in the rhythm of the liturgical text, a patterning that puts praise high on the agenda, the paradise image is one of harmony, where animal and human and life are connected in God. The hint is that powerful need for the Divine One to have companionship and friendship not in hunting or building or subjugating created life, but in that symbol of connection of deep peace in the cool of the garden where one walks with God, enjoying all. That instinctive image is something the liturgy brings out in its texts, a symbol of the Spirit's abiding presence; but there is more. The dissonance of sin, that act of human beings in a world fractured from God, creates a dissonance with life itself. Then animals are for human use, then killing for food becomes an imperative for survival, then blood especially the sacrificial blood of bulls and goats and lambs and turtle doves, becomes a substitute for an eternal covenant a promise that does not require life blood, but this is a human doing, not God. Let us listen to other echoes in the story, in the Noahide story, after that extraordinary time of *Nephelim* and humans, a flood destroys the earth and humanity, but, in contrast to the sullen destruction of human life, all other living species are representatively saved. This is important for us: the Noah story, with the Ark as a constant theme of salvation, taken up in Judaism and then patterned throughout Christian tradition has more animal, reptile and insect in its capacious hold than the humans represented by Noah and his small family. Not only that but the symbol of sacrifice (being sent to find the promised land hope and trust) is given to two forms of birds: 'After forty days Noah opened a window he had made in the ark and sent out a raven, and it kept flying back and forth until the water had dried up from the earth. Then he sent out a dove to see if the water had receded from the surface of the ground. But the dove could find nowhere to perch because there was water over all the surface of the earth; so it returned to Noah in the ark. He reached out his hand and took the dove and brought it back to himself in the ark. He waited seven more days and again sent out the dove from the ark. When the dove returned to him in the evening, there in its beak was a freshly plucked olive leaf! Then Noah knew that the water had receded from the earth. He waited seven more days and sent the dove out again, but this time it did not return to him.' (Gen 8: 6–12)

Ironically the birds are themselves representatives of a much older strata of life than the human being on this planet, namely the dinosaurs, but both are also deeply and symbolically rooted in scriptural, patristic, monastic and liturgical imagery. It is they who identify with hope. In monastic literature following the example of Elijah in the book of Kings (1 kings 17: 2–6), the raven feeds the hermit Paul with bread in the wilderness and brings a double portion when Abba Anthony visits with him![13] The dove is always a sign of the Spirit and of peace. The identification of God's covenant with the rainbow (Gen 9: 12–17) should give us more food for thought, here, at a point in the 'story' where a covenant of eternal dimensions is ratified between God and all forms of life it is the '*charismata*' of other forms of life that provide the material symbol of the promise. If we could only shift our preoccupation from the needs we have and look at the honour given by the Creator to all of creation in this story, we might experience a gratitude for being part of a whole, and the importance of that fact for ourselves, a harmony is being restored, God's Spirit and Word once again reconcile all creation.[14]

We can see definite examples of this story in our worship as John Chrysostom said in his discourse on the feast of the Theophany, for the ark and dove are symbols of salvation not just now but in the future:

> Wherefore also the Spirit descended as a dove, because where there is reconciliation with God—there also is the dove. So also in the ark of Noah the dove brought the branch of olive—a sign of God's love of mankind and of the cessation of the flood. And now in the form of a dove, and not in a body—this particularly deserves to be noted—the Spirit descended, announcing the universal mercy of God and showing with it, that the spiritual man needs to be gentle, simple and innocent, as Christ also says: 'Except ye be converted and become as children, ye shalt not enter into the Heavenly Kingdom' (Matthew 18: 3). But that ark, after the cessation of the flood, remained upon the earth; this ark, after the cessation of wrath, is taken to heaven, and now this Immaculate and Imperishable Body is situated at the right hand of the Father.[15]

V. Sources in the Rites

If we examine the most common of the Eastern Rites, that of the Constantinopolitan-Byzantine Rite, we find various appearances of cosmic harmony written into the textual performance of the rite. Admittedly this liturgical rite has a number of later features, but its origins are ancient. In the blessing of water for baptism that the priest says these words in the long blessing (we have already noted the Noah imagery in the blessing of oil), many other images appear:

> Great are You, O Lord, and wondrous are Your works, and no word will suffice to hymn Your wonders.(x3) For by Your Will have You out of nothingness brought all things into being and by Your power sustain all creation and by Your Providence direct the world. You from the four elements have formed creation and have crowned the cycle of the year with the four seasons; all the spiritual powers tremble before You; the sun praises You; the: moon glorifies You; the stars in their courses meet with You;

> the Light hearkens unto You; the depths shudder at Your presence; the springs of water serve You; You have stretched out the Heavens as a curtain; You have founded the earth upon the waters; You have bounded the sea with sand; You have poured forth the air for breathing; the angelic Powers minister unto You; the Choirs of Archangels worship before You; the many-eyed Cherubim and the six-winged Seraphim, as they stand and fly around You, veil themselves with fear of Your unapproachable Glory; for You, being boundless and beginningless and unutterable, did come down on earth, taking the form of a servant, being made in the likeness of men; for You O Master, through the tenderness of Your Mercy, could not endure the race of men tormented by the devil, but You did come and saved us.[16]

In this context a spirituality inherent in worship emerges clearly in the action of the rite, at this ceremony of baptism, when the catechumens take on a transformative process that ends the tyranny of sin and death by renewing in their deepest selves a 'new' and concrete life, the elements of water and oil are symbolically and metaphorically breathed upon, as a total symbol of that breath of the Spirit. It must be noted that the liturgical text calls these the 'creatures', part of transforming life. (This is also present in formularies from other traditions, the old Roman Rite etc) The representative liturgical pattern of Genesis is replicated and in theological context, the life-giving Spirit effects a sacramental change in all these unitary elements, which together combine to effect the sacramental action. In this sense we discover the oneness of all at the heart of Christian Initiation, but also the connectivity with the Creator who has made all, sustains all and directs all and who in those symbolic words sets the boundaries of creation, which we alas cross over without any regard for consequences (or at least have so frequently done!). Surely if the harmony restored in theological terms which through the words and symbols and explicitly stated in the ritual activity, then it follows that life in Christ, which this part of the rite upholds, is a restoration of cosmos and earth with the heavenly realm, in all its component parts.

As the rite progresses into the baptism itself, the pre-anointing touches all the organs of sense in a particular manner (The servant of God, ______, is anointed with the oil of gladness; in the Name of the Father, and of the Son, and of the Holy Spirit. Amen.), if we examine what is intended here, the restorative action places a huge stress on what is to come, 'for the etc.' and if new sight, hearing touch, feeling are part of the charismata of baptism, then this will also include a conversion, repentance, which itself puts us into a completely new relationship with our world in all its aspects. This might seem to be a laboured point, but if the Christian community takes its role seriously, the Initiation rites should have at their heart a profound meaning for who we are, what we do and our response and place in the world.

That this is not simply figurative can be shown in the corresponding symbol structure taken from the art of the baptistery itself. Excavations in Nazareth uncovered a baptismal mosaic which revealed a complex understanding of the importance of these initiation rites and evidence from fifth century baptisteries such as Ravenna point to a highly developed sense of restorative theology, the theme of 'paradise regained'. In the centre of one of the most famous mosaics, Christ stands naked in the Jordan, whilst about him the mountains lift their hands in expressions of wonder and the cosmos of heaven and earth, angels, saints and apostles in heaven are united together with the event taking place below, there is no absence of animal life either. Those being initiated in to the Holy Mysteries would have no doubt that they were also entering a new promise and the anticipation of 'paradise regained'! Annabel J Wharton has written a very clear analysis of the connection between the baptistery, mosaics and liturgical ritual of the *Neonian*, that is Orthodox, Baptistery in Ravenna, arguing cogently that the idiosyncrasies of the mosaic design is not accidental but a clear exposition of Christian belief and acted as part of the liturgical stage setting for a semi-public ritual performance![17] The landscape of these early Christian mosaics is one we need to return to again and again, for the recurrent theme is that of a new life, a new age come about in Christ.

The late Archimandrite Robert Taft, one of the greatest of present day byzantine liturgical scholars, encapsulates this cosmic approach in the liturgical environment, writing on the *Hagia Sophia* in Constantinople he is particularly clear about its theological importance: 'in no liturgical tradition is liturgical space such an integral part of the liturgy as in the Byzantine, and in no tradition has one edifice played such a decisive role s Justinian's *Hagia Sophia*…(it) was the cathedral church of the city where the Byzantine rite was moulded and celebrated, and where the vision of its meaning, enacted elsewhere on a smaller stage, was determined and kept alive'.[18] The specificity of this architectonic vision, that *Hagia Sophia* somehow is a determinant and prototype for all churches in this tradition does not take away the basic raison d'etre that all this structure and form is for the community that dwells in Gods presence, in this locus, at this liturgy. In the building of *Hagia Sophia* we see the merging of the *domus ecclesia*, that is of buildings constructed for the 'Holy People of God' and multi-purpose in a very direct way, where the sacred is dependent on the gathered actions of the community in whom Christ is present, and the *domus dei*, one which has more overtones of a specific sacral place in 'temple' terms with a stronger emphasis on the presence of the Divine within the building itself. In *Hagia Sophia* and consequent buildings we find mixture of ideas, a fusion of these two theological explanations of presence which makes complete sense in the context of the byzantine liturgy, where heaven and earth connect. The chant of the Cherubic Hymn just before the Great Entrance, where the material gifts of bread and wine are taken from the table of preparation (*Prothesis*) and processed through the assembly before being placed on the Holy Table for consecration, expresses this connectivity very forcefully, for the hymn symbolically incorporates the community gathered for that liturgy, into the very presence of the angels gathered around God's throne.

The choir or people sing: 'We who mystically represent the Cherubim, who sing to the life-giving Trinity the thrice holy hymn, let us now lay aside all earthly care, That we may welcome the King of all invisibly escorted by angel hosts. Alleluia! Alleluia! Alleluia!'[19]

From the rites and rituals of baptism within the context of Initiation, we begin to discover that the call to new life is not something that is to come; it has taken place in Christ for through his resurrection it is part of what we are about. The second strata of this is the appropriation of the Holy Spirit, having died to sin and risen in the waters from death with Him, the neophyte is then illumined by Christ and anointed by the Spirit then led to celebrate and receive the Eucharist. The Eucharist forms a topic which we shall explore later, but in this context we might examine what the first Eucharist received might say, firstly in eastern theological terms, of this new age. Fr Alexander Schmemann puts it this way:

> This life is rooted in the sacraments of baptism and the Eucharist, and the sacraments according to the early Christian understanding are precisely the means of the eschatological life of the Church. They manifest the 'coming aeon' in this world, and they are themselves but the expressions of the Church as the visible sign of the kingdom which is to come, its anticipation in time and history.[20]

The Byzantine Rite states this quite explicitly, somehow the encounter with the creatures of bread and wine, themselves products of a long process (as the Roman offertory puts it 'which earth has given and human hands have made') is not only a sharing in the Divine Life of Christ, it is in itself a transformative act for the believer, we become participators in his heavenly kingdom and shares of the heavenly banquet. Before communion priest and people bow down three times in a posture of humility, whilst praying: 'O God, be propitious to me a sinner, and have mercy on me', it is then that the priest cries out as he lifts the holy bread before giving communion, 'Let us be attentive! Holy things for the Holy' and the people reply, 'One is Holy, One is Lord, Jesus Christ to the glory of God the Father. Amen'.[21] For the believer this is a deep reminder that this communion is a moment of intense communication between the divine person and ourselves, for the human is called by Christ to receive these mysteries contained in ordinary food and drink, but in recieving them, not only shares in the holiness of God but through the grace of the Spirit has the means to become holy themself. That stated, we can extend the metaphor, the Christian by means of this communion then goes out into normal life to extend that communion with others. The implication for a new approach to the environment is clear to see, for in this action all are caught up, all are in fact part of the process of 'becoming' holy, and as we note from the stress on the transformative power of the Spirit in changing these gifts, by extension this holiness is inclusive of all life and gifts on earth. Nothing is excluded except the unrepentant sinner.

VI. Rediscovery and Renewal

However there are other areas of liturgy that help form the structure of a new approach, although I use this term carefully. A new approach is really retrieval, rediscovery, renewal, and opening up of what is already present. It would be difficult to trawl through every liturgical act to discern this focus or restorative justice for creation and animal life, after all our rituals and patterns of worship are for human use, the sacramental act of reconciliation for instance is about the human beings need for reconciliation, no theologian would dare to assume that animals have a capacity for sin or evil activity in the same way as homo sapiens, although conversely our retrograde and appalling actions have often been labelled as 'animal' or 'like the brute beasts' and so forth, terms and nuances that have helped shape the territorial and moral superiority that the human race believes it has over creation. But this distorted approach towards creation still goes on, it is as though we are persistently unable to attribute the evil in life to ourselves and have to demonise external forces such as animals and blame them: rather as the scapegoat in the Old Testament became the vicarious sacrifice for human sinfulness. Then he is to take the two goats and present them before the LORD at the entrance to the tent of meeting. He is to cast lots for the two goats-one lot for the LORD and the other for the scapegoat. Aaron shall bring the goat whose lot falls to the LORD and sacrifice it for a sin offering. But the goat chosen by lot as the scapegoat shall be presented alive before the LORD to be used for making atonement by sending it into the wilderness as a scapegoat (Lev.16. 7–10)

So our continual and appalling catalogue of cruelty towards animals is a painful witness to our lack of understanding and separation from our iconic status as icons of Christ and a continued 'scapegoating' of the innocent. However it is also a sign of a woeful neglect of the deep and rich theology of the Holy Spirit which is present in all creation. Pope John Paul II reiterated this doctrinal dimension in one of his catecheses, reminding people that the Spirit is constant and continual in the work of creation.

'Moreover, the Second Vatican Council, focusing on the Church's mystery and mission in the world, offered this breadth of vision. The Council holds that the Holy Spirit's action cannot be limited to the institutional dimension of the Church, where the Spirit also works in a unique and full manner, but should be recognized outside the visible frontiers of Christ's Body as well (*cf. Gaudium et spes*, n. 22; *Lumen gentium*, n. 16). For its part, the *Catechism of the Catholic Church* recalls with the whole of Tradition: "The Word of God and his Breath are at the origin of the being and life of every creature" (n. 703). And a meaningful text of the Byzantine Liturgy says: "It belongs to the Holy Spirit to rule, sanctify and animate creation, for he is God consubstantial with the Father and the Son.... Power over life pertains to the Spirit, for being God, he preserves creation in the Father through the Son" (*ibid.*). Thus there is no corner of creation and no moment of history in which the Spirit is not at work'.[22]

The Pope is only expressing a deeply held theology the Holy Spirit found in both east and west but one that needs much more attention in the western theological and ecclesial traditions, as the east has always had a much more developed pneumatic theology. In fact the work of the Spirit, who brings us into communion with Christ is one of the consistent hallmarks of its spirituality. There are many gifts to help us, such as teachers from the past. One of the great Christians traditions, that of the Church of the East gives us the work of spiritual teachers such as the Syrian monk and bishop St Isaac of Nineveh (623–700CE) whose homilies are still a source of inspiration to many Christians not least those who are monks and nuns. In one passage Isaac brings together the elements of progression we have already noted in the consideration of the three modalities of presence: Isaac writes

> Until a man has received the Comforter, he has need of the divine Scriptures to imprint the memory of good in his thought, to keep his striving for good constantly renewed by continual reading, and to preserve his soul from sinful ways; for he has

> not yet acquired the power of the Spirit, which repels errors and captures soul-saving memories. When the power of the Spirit has penetrated the power of the soul acting in a man, then, in place of the law of the Scriptures, the commandments of the Spirit take root in the heart and man is secretly taught by the Spirit and needs no help from sensory matter.[23]

Whilst we acknowledge the particular emphasis Isaac is placing on the Spirit as a guide particularly for those who follow the monastic road, nevertheless his implicit connection between the *metanoia* of a Christian beginning in the encounter of Scripture which prepares a person for the engagement with the Holy Spirit, is a solid classical and timeless approach in the spiritual tradition of discernment and growth.

True stewardship is partly the recognition of a deep need for reconciliation not only with my brothers and sisters but with my mother the earth and all God's creatures. Just as tragic is our quest for dominance over nature, the relentless mess that we leave on the face of our planet with exploitation and building at every turn, our approach to sin and repentance should indicate some desire to restore, heal and atone for what we as a race have done, not only knowingly but through ignorance and selfish blindness. As David Tracy puts it in his work, 'The Analogical Imagination':

> Should it be any cause for wonder that those theologies with their major focus on the reality manifestation will be most alive to certain signal realities in the complex phenomenon called 'world'? With their sense for radical giftedness and thereby of God's radical, always-already, loving immanence in all reality, these theologies will be open to a pervasive sense of grace. Nature and body, not only history and spirit, will be felt as graced and thus interpreted in theologies of creation, incarnation and sacrament.[24]

Notes

1. Liturgy of st John Chrysostom, p 22.
2. *Ibid.*
3. Martimort, A.G. 1966. *The Church at Prayer: Vol1 Introduction to the Liturgy*. The University Press, Shannon, Ireland, p 84. See also the Liturgy Constitution of Vatican II, Sacrosanctum Concilium (S.C.2; S.C.7).
4. For a succinct Orthodox exegesis see; Metropolitan Anthony of Sourozh. 2008. *The Living Body of Christ*. Darton Longman and Todd, London. Chapter 1. *The Church, Extension of Incarnation*. pp 3–29.
5. Liturgy of St John Chrysotom, p 16.
6. Ibid, p 28.

7. Schmemann, Fr Alexander. 1991. *Celebrating the Faith: The Creed*. Volume 1. St Vladimir's Seminary Press, New York, NY. p 60, also quoted in the Orthodox Education Service bulletin http://dce.oca.org/assets/templates/bulletin.cfm?mode=html&id=143.
8. See *Dei Verbum* Chapter 1.8, in Walter. M. Abbott, S.J. 1966.*The Documents of Vatican II*. Guild Press, New York. p 116.
9. Brett 2000. p 27.
10. Schmemann, Alexander. *Liturgy and Theology*. The Greek Orthodox Theological Review XVII, no.1 (Spring 1972), Holy Cross School of Theology Brookline, MA. p 90.
11. *Blessing of oil at Baptism. The Office of Holy Baptism.* Service Book of the Holy Orthodox-Catholic Apostolic Church, Third Edition. 1956. Syrian Antiochian Orthodox Church, Brooklyn, NY, p 279.
12. Metropolitan Emilios Timiadis of Calbria. *Restoration and Liberation In and By the Community*. The Greek Orthodox Theological Review XIX, no. 2 (Autumn 1974), Holy Cross School of Theology, Brookline, MA, p 150.
13. Then the word of the Lord came to Elijah: 'Leave here, turn eastward and hide in the Kerith Ravine, east of the Jordan. You will drink from the brook, and I have directed the ravens to supply you with food there.'
    So he did what the Lord had told him. He went to the Kerith Ravine, east of the Jordan, and stayed there. The ravens brought him bread and meat in the morning and bread and meat in the evening, and he drank from the brook.
14. See *Laudato Si*, Chapter One. *What is Happening to Our Common Home?*
15. St John Chrysotom, Discourse on the Day of the Baptism of Christ.
16. *Blessing of Baptismal Water*. Service Book, 1956. p 278.
17. Annabel, J Wharton. 1987. *Ritual and Reconstructed Meanings: The Neonian Baptistry in Ravenna*. Essay in The Art Bulletin. Taylor and Francis. pp 358–375.
18. Taft, Robert F. 1995. *Liturgy in Byzantium and Beyond*. Ashgate Variorum, Farnham.
19. Liturgy of St John Chrysostom, p 38.
20. Fisch, Thomas (ed.). 2003. *Liturgy and Tradition: Theological Reflections of Alexander Schmemann*. St Vladimir's Seminary Press, Crestwood, NY. p 17.
21. Liturgy of St John Chrysostom, p 62.
22. Pope John Paul II. *Catechesis. The Holy Spirit Acts in All Creation and History*. Osservatore Romano. August 12th 1998.
23. Saint Isaac of Syria. *Directions on Spiritual Training*, in E Kabloubovsky and G.E.H. Palmer 1969 op cit. p 242.
24. Tracy, David. 1981. *The Analogical Imagination: Christian Theology and the Culture of Pluralism*.SCM Press Ltd, London. p 436.

· 3 ·

FROM THE EUCHARIST—ASPECTS OF THE DIVINE LITURGY

The central act of historical Christian worship has always been the Eucharist, called also by various historical names including the Byzantine terminology of 'The Divine Liturgy'. From earliest times this act has been a weekly celebration of Church assembled with its Lord, in clear ecclesiological and liturgical terms, and held on the first day of the week, our Sunday! Historical changes over time led to a more frequent celebration of the Eucharistic liturgy such as the normative Sunday celebration, with as many clergy as was possible present, deacons and deaconesses exercising their various ministries, and the presbyters concelebrating with the Bishop, to a more complex development of parochial liturgies with a few (or even one) priest and in later stages of liturgical evolution, particularly within Catholicism, weekday Eucharistic celebrations, especially during the later Middle Ages, became a regular and accepted feature of church life and practice, another significant shift coming from the (originally Cluniac-Benedictine) expansion of private masses in Western Catholic Christendom! Many of these developments did not take place in the east and there were significant differences in theological and liturgical approaches to this sacrament with the west where the eastern church kept to more traditional practices including concelebration by the priests, retention of the distinctive ministerial functions of the 'orders' of clergy that were not duplicated by another, for example the East has never allowed a priest to act in place of a liturgical deacon or wear their vestments as was done in the Catholic 'High Mass'. Other examples are the use of leavened bread, reception of both bread and wine at communion by the faithful and in Great Lent and Holy Week a Eucharistic fast where the liturgy of the pre-sanctified took the place of the Divine Liturgy

What we discover is that in history and actual practice East and West have had different approaches both to the frequency of this celebration and in aspects of theology, but in the main (the exception being with some more radical elements in the Protestant tradition) kept to the ecclesial unity of the Sunday celebration.

The Orthodox theologian, Fr Alexander Schmemann, reflecting on the particular eastern theology of the Eucharist as 'encounter' between God and humankind, makes a clear connection between the exigencies of everyday life for everyday people, the community of the Church and the coming Kingdom; helping us understand that it is through the liturgical 'coming together', the assembly of the community in Christ, that the cosmic and divine presence is manifested in this world in a particular and very real manner by the encounter with the Trinity in its chief act of worship. This means of course a particular view of matter and place, the Christian can never treat the world as something to be used and abused, but something that must be respected, loved and cared for because it is part of the Kingdom, not yet fully realised:

> The mode of presence in this world of the 'world to come', of the Kingdom of God, is the Church—the community of those united to Christ and in Him to one another. And the act by which the Church fulfills that presence, actualizes herself as the new people of God and the Body of Christ, is the 'breaking of bread,' the eucharist by which she ascends to Christ's table in His Kingdom.[1]

This is a theological insight, far from being incidental to any concern we have for this planet and its environment, and it is particularly important that it comes from an Eastern Christian standpoint, it's a reminder that there are alternative theological models of the human's place in creation.

I. The Eucharist, Genesis, Cosmos and World

There are a number of strands within this particular Eucharistic setting of theology and liturgical practice, which can help provide models to focus particular approaches towards the wider questions of creation, environment and sustaining life in differing contexts. As the Eucharist is the 'high' celebration in sacramental worship, its meaning for the traditional Eastern Christian is non-negotiable, that is, whatever the Eucharist means for that community in terms of faith, it is one of the great dominical commands to 'remember and do', and that is why it must be celebrated as it has been very faithfully down the ages:

> The Eucharist has its day, Christians gather together on a *statu-die*—on an established day. We know that the 'Day of the Sun' was not a holy day of rest in either the Jewish or the Roman calendars. Nonetheless the Eucharist 'became so firmly connected with this day' that nothing has ever been able or will be able to undermine this connection.[2]

Whilst there are rich and varied theologies of Eucharist in both east and west, nevertheless from the east we get a different set of perspectives that, in a very real way, moves the whole theological position of the celebration into a dimension of cosmic significance. In this sense there is a deep connectivity for any of our environmental and ecological concerns for creation within the reality of this sacramental celebration, for example in the *Anaphora* (Eucharistic Prayer) of St John Chrysostom in the Byzantine Liturgical tradition, the priest recites these words:

> On the night when He was betrayed, or rather when He gave himself up for the life of the world, he took bread in His holy, pure, and most blameless hands, gave thanks, blessed, sanctified, and gave it to His disciples and Apostles saying, Take eat, this is my body which is broken for you for the forgiveness of sins.[3]

This phraseology takes us out of the confines of any anthropocentricism, for here in the center of the prayer of thanksgiving is a totally cosmic reference, that the Christ gave up his life not only for us and all people, but also 'for the life of the world'. The implication of this is of course to widen Christ's sacrificial gift to the whole of creation and to include creation in the life of God.

What does this imply? Sacramentally it links all life, all creation with the nature of the Triune God. We can discern something of a liturgical origin of this in the tripartite structure of the Genesis I narrative, this is in a worshipping sense the song of creation, the rhythm of the seven day pattern gives us a structure where the Creator 'conceives', so to speak, the Word 'forms' and the Spirit 'gives' life. Now of course, perhaps in strict OT biblical terms, this may not a great exegesis, but it is significant, Walter Bruggemann sees the liturgical influence as part of the 'traditioning process' and in his paper on 'Creation as Temple Building', Jeff Morrow has this to say about Genesis :

> The careful attention to a sevenfold structure indicates that Genesis in its final form is a liturgical text. We may go further and state that, in fact, Genesis 1 reads as a sort of liturgical hymn. On the basis of the heptadic structure, Weinfeld has argued that its *Sitz im Leben* is the liturgy.[4]

These insights are particularly important from the point of view of liturgical theology, especially that influenced by the East where 'tradition' is always part of the context of process and development. We can discern here what I call one of the 'chants' of creation, a rhythmical antiphon in the heart of the liturgical actions of the Church but especially the Eucharist, which is not only found within the scriptural text but also in image, homiletic-text, prayers, actions and symbols, all pointing within the Christian tradition to the community of love, which is the Trinity and in which we share and participate though our worship and prayer.

Perhaps one can refer back to an insight about the nature of God that is deeply rooted in the eastern spiritual tradition, which is 'par excellence' the Church of the Holy Spirit: for each liturgical act begins not only with a prayer to the Holy Spirit, but in particular at the Divine Liturgy of the Eucharist after this prayer the opening and cosmically referenced blessing is sung:

'Blessed is the Kingdom of the Father and the Son and the Holy Spirit, now and always and for ever and ever,' and whilst this is done the Priest lifts up the book of the Gospels and in the signing of the cross over the Holy Table, blesses the world present at the liturgy.[5] This has also been symbolically done by the deacon or priest at the great incensing just before all of this, where as the four corners of the Holy Table are honoured with fragrant smoke and the sanctuary nave and people are incensed, Psalm 50(51) is recited secretly, a psalm of the 're-creation' of a pure heart and the reconciliation of earth with heaven. In such highly charged and symbolic moments, the Word, made incarnate in each of us by our baptism, chrismation and communion with Him also touches the word present at creation. Whilst some may define this as typical over-symbolism from the east, the mythos represented is more than that, it is the real sense and experience of the sacred encapsulating all things but particularly by our actions at 'this moment', our search for the Divine Being acknowledged in an encounter 'now' which is continually being enacted throughout time and in all places.

II. Space and Place: Liturgical Environment Versus Human Environment

One of the most cosmopolitan of the eastern rites is that of the Byzantine tradition, here the layers of theology, symbolism and spirituality intertwine to create a rich tapestry, leading us beyond the purely spiritual connectedness of the community with its Lord, to a more comprehensive overview of the whole world brought together in both a cosmic and physical sense of interactivity. If we consider the eschatological dimension of liturgical action, where because of the resurrection of Christ the new Kingdom is already partially realized, we will also discern that within the structure of the Eucharist itself insights are found that help us to see some very different perspectives on world, humanity and life itself.

But perhaps we should begin with the space of the church itself, this may differ in some aspects from the protestant or catholic understanding of church building, not only because different sacramental, liturgical and ecclesiological theologies operate in them, but because the cultural diversity of communities, and therefore a particular style of adaptation, was acknowledged, respected and utilized from early on. It is important to remember this is very much a developmental approach which is coupled with a basic principle, namely that unity in faith does not suppose uniformity in practice, for instance in the earlier centuries of Christianity the insistence on mystery and awe did not carry the same weight as it did (particularly for the East) in the post-iconoclasm age when celebration of the Eucharist became highly symbolic in approach, one not developed in the same way either in texts, symbols or ritual in those earlier centuries.

III. Starting Point of Space

The starting point of any liturgical space within the Christian context is of course the ecclesia, the community of the Church. Examination of the NT and equivalent textual evidence does not give us any clear picture of a separate building for the cult in its early stages of development. In Acts 3 ff. it is evident that the Temple and Synagogue were still used for worship by the apostles and their followers until events by the Jewish community barred them! Even in the eventual emergence of the distinctive House Church utilized by the people solely for the activities of the cult, adaptability was shown for the locus of particular rituals. For instance in the 'Christian' rite of baptism it is abundantly clear that depending on the situation, which was mainly environmental, different water sources were used, rivers, ponds, lakes, springs, fountains, but also in the case of more desertified ambiances, cisterns or containers.[6] That wonderful handbook for early Syrian Christian practice from the mid second century, the *Didache* has simple directions on the matter, use flowing water, if that can't be found other water, have it cold but if that can't be found then warm and if there isn't much water to be had simply pour.[7]

'The *Didache* teaches that baptism should ideally be administered in "running water" (7.1), by which is meant flowing water, as in springs, streams, rivers and seas. According to Tertullian (De Bapt 4.3) "it makes no difference whether a man be washed in the sea or a pool, a stream or a fount, a lake or a trough". The range of options presumes that baptism often takes place in open bodies of water. The Apostolic Tradition (21, 1–2 as represented by the *Testamentum Domini)* dictates that baptism should take place in "pure and flowing water". A natural body of water—sea, river, lake—would thus seem to have been the most regular and desirable setting for baptism in the early period'.[8]

In the case of the Eucharist (the breaking of bread) until the emergence of the house church in the third century, upper rooms in the homes of Christians probably accommodated the gathered community, what they were doing was a simple adaptation of normal everyday patterns of living. We know that this space was necessary even if direct evidence is sparse simply because of what they were actually doing. The *Didache* for instance, insists on the community gathering on the Lord's day for the breaking of bread (14.1), but also reminds them to 'come together often' (16.)The letter of Barnabas simply commands the community to 'assemble yourselves together' (4.10).

Whilst there is very little description of a place or building, there is a direct context for environmental issues here, one which will give us some theological grounds for evaluating our own considerations of the built environment. In part this comes form a shift in the interpretation of holiness in the sense of sacred space and place, in other words some form of created sacral landscape that restricts the activities and incursions of the community. This is theologically important because we cannot leave out in any equation of religious place the wider implications of our own species, its consistent conquest of the natural planet and its inbuilt need to meddle, sort, shift, change and mess, what one can identify as a need to 'litter' the land with constructions of its own. That some of the buildings and artifacts we have produced down the centuries have an inbuilt worth and aesthetic and historical importance is to be commended, but on the other hand, much of what human beings do in an industrialised setting, with more artificial material produced and therefore less biodegradable, can be very squalid if not downright destructive. In the past our mess and rubbish was far more connected to the rhythms of life and death giving back what had been taken, comprised as it was of natural materials, such a far cry from today's rubbish, where artificial materials such as plastic create havoc with the different environments, on land, in space and under the oceans!

In previous ages smaller numbers of the human family ensured that we could not over-farm the resources of the natural world, although there is also other evidence of ecological warfare, evidence shows that as some civilizations developed others declined because of our destructive tendencies to exterminate other life, and also create what is called 'land crisis'. A report by Chinese Economists on 'Global Environmental Competitiveness' makes the clear point that from about 8,000 years ago, when farming became widespread; 'expansion of farming land destroyed vegetative cover, forests were laid down and grasslands were cultivated followed by soil erosion and desertification… These were all destructive', and have led to what is still a major environmental issue of land crisis throughout the world![9] This can only get worse with the massive increase of the human population. We are also not particularly good at clearing up our mess! There is a huge chasm between the archeological detritus of the medieval rubbish dump, a resource for investigating our ancestors, where it is evident that most of what has been thrown away disintegrates in time, to the vast landfills of the late twentieth and twenty first-century, some of which are highly unstable places, not to mention the detritus we have left orbiting our planet in space! Technological advancement has not always meant good environmental and ecological progress!

It is fortunate that good ecological and environmental awareness is tackling such issues, but it is a terrible indictment of our first world profligacy, that we bury and destroy much of what the second and third world would see as usable materials fit for endless recycling. In this sense a theology of recycling drawn from the deeply liturgical theology of the east may have a prominent part to play in creating a balanced spiritual viewpoint that has a much more humble approach and does not put humankind at the centre of everything. If for instance, we keep returning to a better and more focussed understanding of human stewardship, in which the spiritual gifts of loving service and humbleness of heart take centre place, then the words taken from I Chronicles used in the Eucharistic prayer of Chrysostom: 'For all things come from you, and of our own have we given to you', can be seen as a divine mandate for creative and constructive recycling.

It is into this scenario that the Eastern Christian theology of place and space creates a matrix for confronting and challenging as well as restoring some sense of good order! How? Returning to that early sense of place and space, eastern Christians can acknowledge a biblical and theological world-view which acknowledges the world as implicitly belonging to the Creator, but also as part of the human setting, to which they belong. For the Christian community the Christ event did not change that but transformed it. In the Risen Christ all things are made new, not abandoned or rejected like rubbish but reformed, recycled so to speak! The ultimate abandonment of death is itself changed and challenged to become for us all, hope in resurrection but also resting in peace. In the ancient world, the sacred place spoke of the separation of a numinous, different world from that of the ordinary, one usually entered by death or at least by some form of death like experience. Key to this experience is a sense of liminality, our ancestors used natural objects like stones or hills or rivers to identify sacred places, but they also created and built spaces, with quite well defined boundary areas, where this world could meet and interact with the 'other', these buildings we call temples, places to house the God or Gods, dwellings where the devotee either in person or through the intermediary of a religious figure, connects with the Divine Realm. Ancient Egypt, Persia, China, Greece and Rome, for instance, still provide us with many examples of such places. There are also many pre-historical places that suggest separation and ritual observance of a god like figure or figures, though in these cases documentary evidence is scarce and conjecture often fills the gaps with folklore, yet even that has a source in something! .

Nevertheless, all these sites tell us of a human need to interpret and at least connect with something outside of the normative experiences of life, or perhaps to make sense of those times when the human person is in danger, the moment of birth, adolescence, sickness, adulthood, old age, dying. It is interesting to note that the sites of ancient Britain such as Stonehenge and Avebury were intimately linked to the ancestors, and the barrows of the dead that surrounded these sites were not simply status burials but had a commemorative place in the landscape of place and memory. That too transferred itself across the religious traditions of the ancient world. Jerusalem for the Jews became the ultimate holy city where God dwelt in his temple, it became the center of worship and sacrifice, at its center was the Holy of Holies, so sacred that nobody except the High Priest could enter, a space redolent with God's majestic presence, containing the Ark of the Covenant (1 Kings 8 ff) but no image of the unseen God, because this was where the presence of the unseen

God could be located at particular moments. Yet in the prayer of Solomon at the dedication of the Temple we get an echo of something that enables us to shift from place to person:

> But will God really dwell on earth? The heavens, even the highest heaven, cannot contain you. How much less this temple I have built! Yet give attention to your servant's prayer and his plea for mercy, Lord my God. Hear the cry and the prayer your servant is praying in your presence this day. May your eyes be open towards this temple night and day, this place of which you said, 'My Name shall be there,' so that you will hear the prayer your servant prays towards this place. Hear the supplications of your servant and of your people Israel when they pray towards this place. Hear from heaven, your dwelling place, and when you hear, forgive. (1 Kings 8: 27–30)

It is clear even from this very sacral type of prayer, that the God of Solomon is not envisaged as being in any way confined or 'permanently present' in this place, in fact the building is only a marker for encounter with the Divine One who is greater than anything we can make, but who is in a relationship with His people and when prayer is directed towards this locus will answer them and forgive! The key here is not temple of stones but a living people connected through this particular place.

IV. A Shift From Place to Personalized Space

This shift clearly takes place within the Christian tradition. The answer to Solomon's question is made abundantly clear in Christ. Early Christian theology did not pick on this sacral, temple aspect of place at all, a subtle and important change occurs with the presence of Jesus, for in him a 'replacement' sense of sacredness emerges, that of the holiness of God's presence in all life, but in a particular manner with life in the 'community gathered in His name'. The emphasis on relational sacredness is underscored by and through the sacrament of Initiation, water baptism, dying and rising in Christ and gift of the Spirit in Chrismation, mark out the neophyte as a child of God, sharing in God's life totally. Grace is internally located in the persons themselves who become metaphorically speaking, 'living' temples of the Holy Spirit and the connective Body of Christ. It is through the formalized liturgy that we see how location of place and space moves, the temple model of stasis becomes moving personalized space of relational encounter, the 'living temple' in Christianity.[10] It is clearly understood early on, in the John's Gospel, in the story of the cleansing of the temple, Jesus prophesies its destruction and his rebuilding of it, but as is pointed out: 'But the temple he had spoken of was his body. After he was raised from the dead, his disciples recalled what he had said. Then they believed the Scripture and the words he had spoken' (Jn 2: 22).

The Incarnation of Christ means that sacredness is different for the Christian community for them, the world and all it contains is redeemed and blessed, the separation of profane and sacred is no longer applicable. For where the community is there is God, where two or three gather, there the presence is, all becomes the place of sacred encounter. As if to emphasize this, the destruction of the Temple in 70CE passes the Christian community by, all references to Jerusalem and its temple become linked to the vision of the 'new' heavenly city' as we can detect both in later liturgical texts and the Book of Revelation. Moreover there is also that transposition of sacred place to something more personalized. In the letter of Barnabas we find some of that theology of living temple discovered in writings such as I Peter 2: 5; 'You also, like living stones, are being built into a spiritual house', or Paul's image as the body as 'living temple' in I Corinthians 3: 16 .The Letter of Barnabas states that the Christian is the living temple and their task is 'to become a temple perfect for God' (4: 17) and the locus for that is personal not spatial, 'For a holy temple unto the Lord, my brethren, is the abode of our heart' (6: 15).

This shift means that the early liturgical acts we now understand as our liturgies and sacraments, took place not in a temple setting with the hierarchy of intermediaries, but primarily in a domestic place, that of a community enterprise created from local buildings.[11] In part there had been the influence of the synagogue tradition, where a building was used for a multiplicity of purposes as was the House Church, and in spatial terms there was one legacy of two particular focal points, chair of Moses and place for the Word, and which eventually gives to the Christian Liturgy the bishop's *cathedra* and not only the distinctive space for the Liturgy of the Word, but also its structure. This can be clearly seen in the ancient Syrian tradition where the space known as the *Bema* was a completely separate construction right in the midst of the nave. Part of early liturgical action also took place in a more 'domestic' setting with the Upper Room being used for the breaking of the bread, the *agape* meals and other services such as preaching. Early Christian art shows this in some of its early frescoes found in the Roman Catacombs, and we know from comments such as Paul's about the Corinthian meal that the isolation of meal from eucharist took some time, which means that the disposition of elements into purely liturgical spaces evolved gradually. (I Cor 11: 17–34)

In what way does this affect a theology of Eucharist and of our concern for wider environmental issues, especially from an eastern perspective? The answer is of course that the origin, the source, always links to that progressive aspect of 'tradition' backwards and forwards in time, for what is done now has a direct link to the UR event, but in liturgical action, '*anamnesis*' in its theological sense, what we can call, 'remembering the now', brings us to the original event in a *kairos* moment so that for example, we confidently proclaim in our paschal celebrations that NOW 'Christ is Risen he is Risen Indeed'. For the eastern churches their place of celebration of the Eucharist, that is its liturgical environment, speaks through the liturgical architecture, art and symbol about the theological world view of the community gathered together. In this context there are stratas of existence, for instance the interaction of community and God in word and sacrament within the context of place and living encounter. Convened as the assembly of the Lord, the community in its worship is the actual locus of direct interplay between Christ, Spirit, Father and each person present which at the end of the celebration is commissioned to move out from that context and bring blessing and grace into the wider place of house, home and natural world.

But there is more, we shall of course examine the symbols used in the Eucharist, nevertheless all these elements have a direct connection to creation and ecology, there is also that basic pattern of death and life recycled in different activities, in the Eucharist the act of eating and drinking itself, for nothing is wasted, all is touched with the presence of the triune God and shared with us! Alexander Schmemann in his book, The World as Sacrament, reiterates this connectivity:

> ...the world is meaningful only when it is the 'Sacrament' of God's presence. Things created merely as things in themselves destroy themselves because only in God have they any life. The world of nature cut off from the source of life, is a dying world, it is communion with death. Food itself is dead, it is life that has died and it must be kept in a refrigerator like a corpse.[12]

What Schmemann's comment gives us is that typically Byzantine insight of the whole world caught up in a cosmic liturgy, where everything not only has the potential to be a symbol of '*hierophany*' but can also be deeply embedded in the very life of God. 'All that exists is God's gift to man(*sic*), and it all exists to make God known to man, to make man's life communion with God'.[13]

The development of church buildings took several centuries to accomplish in any hieratic and sacral sense ,it evolved from a very domestic setting where cult focused on the ecclesial community and its needs, for example the Christian building at Dura Europos in Syria (240CE-250CE) is a good example of the *domus ecclesia*, the house church or as one can put it more acceptably, the 'house of the Church'.[14] But though evolution took time, it is also clear from these sort of buildings and what is found in their layout, that a complex (but not mysterious) symbol structure existed. In the Dura house-church, the baptistery walls and the 'font' itself were found to be covered with frescoes of scriptural images though many of these were badly damaged. But by good fortune, in the Synagogue excavations better preserved scriptural frescoes were discovered, possibly painted by the same artists, which helps fill in missing details of subject matter, style and form. Here we can discover an art form 'shared' between Christianity and Judaism, the chosen depictions placed in their context of a particular liturgical setting (synagogue hall and baptistery)made it clear how we are to interpret the different scenes portrayed in the frescoes. There is a clear iconographical programme and in the case of the Christian building the various iconic scenes have a symbolic connection to the meaning of baptism in the story of salvation! Does this early art teach us anything about environment and ecology? Yes, in the sense that it reiterates the connectivity between the divine world and our own, and avoids a grandiosity that suggests the art is done to enhance human superiority, rather the story of salvation keeps us in our place as a servant of world, which is what stewardship on behalf of God is all about!

The next transition was the transition from a multi purpose '*domus ecclesia*' to the more powerful public and cultic buildings took place when Christianity first became legal under Constantine after 315 CE, then 'the' main religion of the Empire, in terms of building, two distinctive styles emerged over time. The Basilican style, a multi-purpose building, rectangular, often with aisles, and a semi-circular apse at one end, found in both eastern and western sections of the empire, and also a cross shaped building with a central dome that became the main type of structure in the east. Neither of these two styles were, at first, invested with sacred symbolism although there is evidence that the developing theology of the Sacraments such as baptism and eucharist begins to shape how the interior was designed. As with ordinary human life and its daily needs, ritual performance in the Christian setting demands its necessary liturgical and functional objects, a table for the celebration of communion, a reading desk for the scriptures, homily and intercessions, a water bath for the immersion of those being baptised and in the big churches a *cathedra* or seat for the bishop whose teaching authority is exemplified by that chair. In one sense the early form of Christian building is shaped by the practical needs of the community, not only in worship.

In ecological terms it is a multi-purpose building, blending in with others around it, and in no sense destroying landscape, rather enhancing it. As bigger buildings emerge over the later centuries, the place of these buildings and what they symbolize does provide a clear architectonic pointer to a relationship between the divine and earthly life, in which glory, praise, beauty and symbolism are brought into the place of normal human habitation or in pilgrimage sites act as places of deep encounter. At the heart of this all is a deep understanding that Christian eschatology embraces the material as well as the spiritual world, all places and spaces where the community gather in the name of Christ, particularly the Eucharist give to the believing community a taste, a sign and an image of the Kingdom of God. The Eastern Church has always had a glorious eclecticism in its theology, joining both east and west. One could describe it as a theology without pretense, rooted in patristic thought but very much located in the lived experience of the Church's life, particularly the liturgy and the spirituality that comes from this. It is because it is a theology *in via* that its insights reconnect us with deeper truths about our place in the world that is whole![15]

We should not underestimate the notion of the world as a sanctuary, which is mirrored in the earthly construction of the buildings of the Church. In a relatively late description in one of his homilies, the 5th century Nestorian teacher, Narsai, describes a connection between building and heaven that shows a complex liturgical and ecological understanding within the liturgical theology of the Syrian Christian community; that this earth in some way mirrors the heavenly realm:

> A holy temple the Creator built for them of earth, that in it they might offer the worship of love spiritually. A holy temple and a holy of holies He adorned, He fashioned: a sanctuary on earth and a holy of holies in the heavens above.[16]

Though this is a homily on the nature of the Church and priesthood it does contain within it the progressive and gradual shift of emphasis (although this never disappears) of the whole baptized community to a more connective holiness mediated by Christ's representatives, described here as priests. The gradual increase of imagery used in the main building of the church reveals this gradual re-appropriation of sacral imagery and language, but it never fully loses the sense that all of this can only be built and used for the people rooted in a present world. Bernard Cooke has this helpful insight to help us understand the situational theology of this process:

Christians as communities of believers, visible witnesses in space and time to their belief in the presence of Christ, 'situate' the risen Christ in space and time. As risen, Christians believe that he is beyond our set of dimensions and our historical happenings, but he continues to be part of our history because his presence occurs in Christians faith which is in space and time. Christ's presence, as all personal presence, is an *active* reality; it is his self-communicating with those who 'hear' him. His self-gift to believers is a transforming influence in their lives that enables them to change the world in which they live and thereby co-create with him the Kingdom of God.[17]

V. Modes of Presence

It is perhaps in this sense that we understand the building used for worship, particularly for the eucharist has a particular mode of presence in the landscape of architecture and why it is imperative that religious communities take seriously the impact and statement their buildings make on and for the environment. Maybe we should take note of Robert Taft's insight about the way liturgical space plays a huge part in the visible theology of the byzantine tradition, and why in the context of ecology, environment and human footprint, the past might give some theological matrix for our present and future, for in this tradition, the church building is not simply utilitarian, nor is it a Temple, it is far more complex as a place where many realities interact and as such has to take note of the different balances between pastoral care of people, communal activity of worship, private place for devotional and intimate prayer, meeting place and so forth. In this setting the east reminds our world very forcibly that building is also enhancement of human world but also a balanced harmony with other elements, for this is not simply any building it is one in which the Divine connects with ALL creation and as such it is a sanctuary place, a space full of meaning. Taft writes of *Hagia Sophia*:

> In no liturgical tradition is liturgical space such an integral part of the liturgy as in the Byzantine, and in no tradition has one edifice played such a decisive role as Justinian's Hagia Sophia…(it) was the cathedral church of the city where the Byzantine rite was moulded and celebrated, and where the vision of its meaning, enacted elsewhere on a smaller stage, was determined and kept alive.[18]

In this sense the specificity of the vision concerns the basic theological insight that the community is the true dwelling of God's presence, but where the material world of creation is also drawn into the symbol of cosmic temple, the setting for the connection between heaven and earth. The eastern church did not invent this image of church as cosmos, it emerged from scripture and the enacted theology of the liturgy itself, in the scattered imagery of byzantine church buildings we find the constant reference in iconography to the upper reach, that of heaven, the thrones and cherubim, the *Pantocrator* and saints to the lower reach, where human life exists. There is here something of the sense of joy that our world needs so badly, but a joy rooted in divine love. When the ambassadors of Kiev came to seek a religion in 987CE, they discovered in *Hagia Sophia* something that almost defied words, it was not simply the impact of the architectural greatness of Justinian's great church, it was also the combination of building, symbol and liturgy together, the liturgical space itself created the impression. Prince Vladimir's account is justly famous but deserves repetition here:

> We knew not whether we were in heaven or on earth. For on earth there is no such beauty, and we are at a loss how to describe it. We only know that God dwells there among men….[19]

This is of course only one viewpoint amongst many, for some the liturgical and architectural setting of Constantinople might seem excessive and other Oriental Eastern Churches do not have the iconographical traditions of the Byzantine communities, but nevertheless they would all understand the setting for the Eucharist as an environment of holiness in which the overarching mode of presence and action of the Holy Spirit is stressed constantly. We need to be careful about creating an artificial sense of apartness and distinction between the sacred and the secular especially in reference to the use and meaning of the church building. The different emphasis found especially in the byzantine east of connectivity and presence between the divine and human realms, the insistence of the transformation of matter as a gift of God through the Spirit, all rooted so deeply in a theology of incarnation, helps us see that the 'useless' beauty of the building, its space, loci, symbols and rites allows us to reconnect with all creation and allows us to express and experience a unity with God in the life of the world! It is in this space where the faithful actively remember the gift of Christ given to them but also encounter him in humbleness of heart through communion, 'It is communion that makes things "be"; nothing exists without it, not even God'.[20] For Christian perception of sacred space the building stands as a modality of personalization in which secular and sacred become harmonized through the epicletic action of the Holy Spirit.

VI. Spirit, Joy and Trinity

This developed pneumatology is one of the big differences between east and west, because it is the transforming action of the Spirit that enacts change, something that links us to the energy of God in the whole of creation itself. It is how holiness enters people and places, at all times and situations, here and everywhere. We enter this through the life-giving Sacraments of Initiation, by water in baptism, by anointing in Chrismation and by communion in the Holy Eucharist, but our chrismation (confirmation) is our own personal Pentecost, when we enter the new life of the Holy Spirit the life of the Church and our world itself: 'Confirmation is the opening of man(*sic*) to the wholeness of divine creation, it is the true catholicity of life. This is the "wind", the *ruah* of God entering our life, embracing it with fire and love, making us available for divine action, filling everything with joy and hope'.[21]

It is also the expression of joy. In the sequence of the liturgical cycle we sense that the proclamation of joy begins with the nativity of Christ, where 'tidings' of great joy are proclaimed to the outcast shepherds (Luke 2: 10) and where the Gospel narrative ends with the Ascension where the disciples worshipped Christ and returned to Jerusalem with great joy (Luke 14: 52) This is a joy that Alexander Schmemann maintains we must recover if we are to hope for a better world. For those interested in creation it is a joy in all things that really matters. Where do we find this joy? Again we return to the insight of the Byzantine East, the perspective of the Eucharist as something transformative… 'and we have no other means of entering into this joy, no way of understanding it, except through the one action which from the beginning has been for the Church both the source and fulfillment of joy, the very sacrament of joy, the Eucharist'.[22]

In this sense the enriching Eucharistic theology of one section of Christianity can leach out into a wider context. This of course leads us back to 'communion', the essence of the Eucharist itself. Saint Basil developed a particular theology of the Holy Spirit which developed a dynamic Trinitarian viewpoint, he understands the Trinity as a profound communion of being and with his other Cappadocian collaborators develops a powerful way of speaking about person:

> This new concept of person is radically relational. The Trinitarian Persons are always Persons in relationship. With the Cappadocian theology, God is understood as persons-in-Communion. This theology puts persons and relationships as the origin and goal of the created universe. Everything that exists springs from the divine communion and will find its fulfillment only in this communion.[23]

This of course takes us back to the liturgical action of the community called Church where the theology of activity and communion is expressed in word, gesture, symbol and rite. In essence the eastern understanding of Eucharist gives us a deeply enriching mission, to be-in-communion with life itself, all life not just humanity and the life of the world to come! 'By participating in His (Christ's) Supper Christians receive into themselves His life and His Kingdom, i.e. the New Life and the New Aeon. In other words the eschatology of the Eucharist is not "world renouncing"... but above all the affirmation of the reality, the certainty and the presence of the Kingdom of Christ which is "within"...but which will be manifest in all glory at the end of "this world".'[24] That perhaps is one of the most important spiritual and ecological insights of the East! The image of joy at the heart of this relationship is fundamental to a future worldview, for too often the present age is mired in negativity. The sense of the Eucharist being also a journey in reality and faith is key to this shift towards 'joy'.

Perhaps this is where the Byzantine symbol of the Divine Liturgy being also journey or procession is important. The very opening words, 'Blessed is the Kingdom of the Father and the Son and Holy Spirit' announce the destination of the whole of creation, it is towards fulfillment, not symbolically but in reality. By assenting to this Doxological blessing, using the word Amen, we agree: '(Amen) It is one of the most important words in the world, for it expresses the agreement of the Church to follow Christ in his ascension to the Father, to make his ascension the destiny of man(*sic*)'.[25] The beauty of the Eucharistic liturgy, is useless in the sense that it has no other purpose except in terms of celebrating in rite an expression of joy and love for God. This joyful love connects us with the 'uselessness' inherent in creation, reminding us perhaps of a wider scale of values than the utilitarian or human centered need. It is the foolishness of God greater than human wisdom, which we also celebrate with thanksgiving for all that it offers and communicates to us. The movement of the Divine Liturgy is a passage from the old to the new, from 'this world' to the 'world to come'. In the tradition of Byzantine (and other Eastern) liturgical art, vestments, objects, and text the glimpses of the heavenly an cosmic realm continually intrude. For instance the insertion of the *Trisagion*, '*Agios O Theos*, Holy is God!' what is called in English the 'thrice holy' invocation, sung before the liturgy of the word is not simply our response, but an angelic hymn, literally sung and taken up from us by the angels. The priest's prayer that accompanies it states this clearly and links that incarnational message with our own *theosis*:

'O Holy God, who restest in thy Holy Place; who art hymned by the Seraphim with the thrice-holy cry, and glorified by the Cherubim, and worshipped by every heavenly power; Whom out of nothing hast brought all things into being; who hast created man after thine own image and likeness and hast adorned him with thine every gift'.[26] Fanciful? Maybe to some western minds, but that is why we need such eastern insights to forcefully push us out of the mundane framework we craft for ourselves and to learn a set of different priorities and cultural ethos. There are numerous examples of a greater cosmic dimension here, even the greetings from priest to people such as, 'Peace be with you' or 'let us pray to the Lord', echo this deep connectivity, the greeting is from Christ to us all, for in Christ humanity and life returns to God and in Christ expressed particularly by the various modalities of his presence in the Eucharist, Christ returns again to us. We might determine this as a link to Christ's universal role, anticipated and celebrated in the action of the Eucharist.

VII. Word and Sacrament

Though there have always been two basic structures contained within any traditional Eucharistic celebration in the ancient churches, that is the Liturgy of Word and of the Sacrament, they are not seen as two different entities, Schmemann considers the sacrament as a 'manifestation' of the Word.[27] In physical terms we can see this in the attention and honour paid to the Book of the Gospels, not as an object in itself but as the presence of Christ the proclaimed word. In certain traditions the place for this liturgy is distinctive and marked out by definitive physical focal points with their own boundaries as in the Syrian '*bema*' or the byzantine place of the '*ambon*'. Robert Taft points out that according to Sozomen, in the great church of *Hagia Sophia*, John Chrysostom preached surrounded by crowds, so he placed himself on the reader's *ambo* in the midst of all and taught them there. Paulus Silentarius in his dedication hymn for the church, mentions the immense size of the *ambo* and the fact that in that space it was the first liturgical item seen by the people on entering the main worship space.[28] In the eastern tradition it is from this place of the *ambon* or *bema* that the Word is brought to and taken from, dependent of course on its architectural setting. Here too the scriptures are proclaimed, the Gospel read, and the homily is preached, but in Byzantine Great Entrance it also marks a position where the movement of the gifts from the table of preparation, through the whole church and the royal doors towards the place of offering, the Holy

Table itself, is for a liturgical moment, seen as the transitional point of complete transformation. Not only is the *ambon* at this moment an *omphalos*, a still turning point in the liturgical world, it is an echo of the word of creation that brings us to life. In a theological sense this is also a movement from the creation to redemption, the word was present at creation and incarnate in Christ who brings about redemption, all this is caught up and reflected in the liturgy. The prayer recited by the priest at the liturgy of the presanctified amply reflects this; 'Thus your Word having his abode in us and walking in our midst, O Lord, we become the temple of your all-holy and adorable spirit, being delivered from every wile of the devil working in our acts, words or thoughts; and we shall obtain the blessings you promised us, together with all the saints who pleased you since time began'.[29] In normal liturgies the priest prays, 'May the Lord God remember us all in his Kingdom, now and unto the ages of ages'. In this way we are called to exercise the vocation of what the Greek-Orthodox theologian, Metropolitan John Zizioulas terms, 'priests of creation' which above all for the eastern Christian finds its fullest expression in the Eucharist.[30] The phrase 'priest of creation' is very helpful when we are examining relationships of humanity to other forms of life. I have already suggested that one of the better models for that difficult word of stewardship is to be more a 'friend and carer' and in that sense modify metaphors to encompass the praxis of Christ's service for and to others.

In his book 'Ecology at the Heart of Faith', Denis Edwards sums up the particular orthodox viewpoint of this concept:

> When humans come to the Eucharist, they bring the fruits of creation, and in some way the whole creation, to the Eucharistic table. In the Eucharist creation is lifted up to God in offering and thanksgiving. In the East, the central Eucharistic prayer is known as the Anaphora, a word that means the lifting-up. The gifts of creation are lifted up to God, and the Spirit is invoked to transform the gifts of creation and the assembled community into the body of Christ.[31]

It is in the next stage of the liturgy, the *Anaphora*, that this movement takes form in action, the gifts are brought, placed, transformed and returned in communion, but this is not done by any consecratory action of a priest, rather for the eastern Christian this is particularly the work of the 'Life Giving Spirit' through the epiclesis where the prayer makes explicit the transformation of created matter:

> send down thy Holy Spirit upon us and upon these gifts here spread forth…that those who shall partake thereof may be unto cleansing of soul, unto the remission of sins, unto the communion of thy Holy Spirit, unto the fulfillment of the Kingdom of Heaven.[32]

Having then received these gifts we return to the world fed by these heavenly gifts.

How then does all of this affect our understanding of creation and ecological issues? In one sense it is very simple, the eastern approach to Eucharist profoundly challenges our twenty-first century concerns. In so many ways those of us *homo-sapiens* who live in developed countries seem to have lost part of our ability to see life as connective with and in a much wider, bigger, order of things and reality. Partly this is because of shortsightedness perhaps due to selfishness? In our consumer driven world we tend not to move beyond the confines of an increasingly egotistical focus of dreams and desires. As Paul Brockelman so aptly puts it:

> By seeing nature and the entire universe as a 'stuff' put here for our endless productive transformation and use, we have reduced reality to a mere extrinsic value for us. It is no longer encountered as intrinsically valuable in itself. As a consequence, we have lost any sense of belonging to a larger and more significant drama and reality.[33]

This is a highly significant point, the Christian community has not always had a good track record in dealing with nature beyond the human setting of things. It is true that people remind us of a world in pain and suffering, movements concerned with ecology have repeatedly brought the bigger picture into focus and initiatives have been constantly set up. But in terms of the Christian Eucharist, the great central act of worship, there are a number of issues that need to be brought into the open, issues which the eastern perspective can help conceptualise in a different way. We still have to ask the question, in the light of Eucharistic understanding what is the correct relationship we humans have with nature, earth and the cosmos? Brockelman suggests that 'human beings are the reality of nature now conscious of itself', but in the light of scientific and biological advances can we exclude other living things from any definition of conscious reality?[34] Even an attempt to define human life as one capable of living what we call a 'spiritual life' can be open to question. Perhaps a different perspective helps, certainly the dimension of all encompassing creation helps enlarge the horizon. In one of his Easter Sermons, Gregory of Nyssa talks about Holy Pascha, the great feast of the resurrection, which is at the heart of every Eucharistic celebration, as a festival where the whole world is gathered as 'one household'. Here he makes little division between humanity, animal life and indeed all creation. For him all life is brought to its complete fulfillment in the resurrection, here he is giving us some rather graphic natural images of the resurrection transcending not only time and space but creature as well!

> Because of this day, a part of the body which flesh-eating birds devoured thousands of years ago will not be found wanting, and what sharks and dogs and sea creatures fed on will rise together with the reviving man, and what fire burnt up and worm consumed in graves, and in short all bodies which destruction annihilated since creation, will be yielded without defect and perfect from the ground.[35]

There is something rather robust and refreshing about this attitude, alright we might make some allowances for homiletic license, but the fact Gregory uses these types of images as examples of the resurrection immediately takes us out of a much narrower definition of salvation into a comprehensive and all encompassing Pascha, where the Passover of Christ celebrated on this feast and at each Eucharist is the eschatological hope of the resurrection of all life in the Triune God.

The many different aspects of eastern euchology, sacrifice, thanksgiving, memorial that I have left out is not for any theological reason but simply because I wish to concentrate on insights that speak to the issues of environment and ecology and might help us shift our own perceptions and change our ways of acting. But there is one aspect of eastern eucharistic theology which pushes us into a commitment with our world because of who we are with God. The union of each believer with Christ (and in Orthodox theology, the Trinity) is not symbolic but genuine and integral to our life. Cyril of Jerusalem expresses clearly the eastern belief in the presence of Christ in the eucharistic elements:

> With full assurance let us partake of the Body and Blood of Christ; for in the symbol of bread is given to you his Body, and in the figure of wine his Blood, so that you, by partaking of the Body and Blood of Christ, may be made of the same body and the same blood with him. For thus we come to bear Christ in us, because his Body and Blood are distributed through our members. This is how, according to the blessed Peter, we became partakers in the divine nature....[36]

The very act of communion itself when we receive the body and blood of the Lord is a reciprocal encounter, we become part of his body, the many become one as Schmemann puts it so well, 'a manifestation, of the realities of which the saving work of Christ is comprised.'[37] The Eastern Church places such a huge stress on the importance and salvific effect of the Eucharist, calling it the Sacrament of Unity, but this transformation is achieved through the work of the Holy Spirit because we come into communion with Christ and the Father through the communion of the Holy Spirit. At the end of the *Anaphora* is the *epiclesis*, the invocation of the life—giving spirit, an organic prayer in which remembering becomes transformation. Ordinary matter, the created and constructed material of life, bread and wine, has been reformed by human hands and offered to God, 'We offer You your own, from what is your own, in all and for the sake of all'.[38] In return, as Eastern theology teaches. there is now a manifestation of real communion between God, humanity and all creation. Here is the prayer in its entirety:

> Moreover, we offer You this spiritual and unbloody worship, and we ask and pray and entreat: send down your Holy Spirit upon us and upon these gifts here offered.
>
> And make this bread + the precious body of your Christ. Amen.
>
> And that which is in this chalice + the precious blood of your Christ. Amen.
>
> Changing them+ by your Holy Spirit. Amen. Amen. Amen.[39]

Schmemann reminds us that the whole Liturgy of the Eucharist is concerned with bread and wine, created material 'food that God created from the beginning as life'. (Gn 1: 29) For our purpose this underpins one of the modalities of ecological encounter that takes us away from a purely rational and subjective approach to environmental issues.[40] Here in this prayer the bread is transformed by the Spirit into the 'bread of life' not death, the food of immortality, the significance of this is an essential factor to any theological insight which helps recover a balance in order to change our present human approaches to world and creation. 'Christ is life and therefore food. He offered this life in sacrifice "on behalf of all and for all", in order that we might become communicants in his own life, the new life of the new creation, and that we might manifest him as his body'.[41] On behalf of all, the Christian community accepts this gift by assenting 'Amen'. Despite all our worst fears, in this action a great hope is sustained!

Notes

1. Fisch, Thomas (ed.). 2003. *Liturgy and Tradition: Theological reflections of Alexander Schmemann*. St Vladimir's Seminary Press, Crestwood, NY. pp 126–127.
2. Schmemann, Alexander. 1966. *Introduction to Liturgical Theology*. The Faith Press Ltd, London. p 63.
3. Liturgy of Saint John Chrysostom. 2004. Librarie St Paul-Alleluia Press, Jounieh. p 46.
4. See: Bruggemann, Walter. 2003. *An Introduction to the Old Testament: The Canon and Christian Imagination*. Westminster John Knox Press, Louisville, KY, Morrow, Jeff. *Creation as Temple Building and Work as Liturgy in Genesis 1–3*. Journal of the Orthodox Center for the Advancement of Biblical Studies 2, no. 1 (2009), p 2.
5. Liturgy of St John Chrysostom, p 12.

6. See Yarnold, Edward SJ. 1994. *The Awe-Inspiring Rites of Initiation*. Liturgical Press, Collegeville, MN.
7. See Didache 7.1. *On Baptism*.
8. Adams, Edward. 2013. *The Earliest Christian Meeting Places: Almost Exclusively Houses?* T&T Clark, Bloomsbury. p 188.
9. Jianping, Li (ed.). 2014. *Report on Global Environmental Competitiveness*. Springer Open.
10. See Gibbons, Robin. 2006. *House of God, House of People of God*. SPCK, London.
11. See Adams 2013.
12. Schmemann, Alexander. 1974. *The World as Sacrament*. Darton Longman and Todd, London. p 18.
13. Schmemann 1974, p 14.
14. See: Baird, J.A. *Photographing Dura-Europos,1928–1937. An Archeology of the Archive*. American Journal of Archeology 115, no. 3 (2011), pp 427–446. See also http://artgallery.yale.edu/online-feature/dura-europos-excavating-antiquity.
Carl Kraeling, H., and Bradford Welles, C. (eds.). 1967. *The Excavations at Dura-Europos, Final Report VIII, Part II, The Christian Building*. Dura-Europos Publications, New Haven, CT, Distributed by J.J. Augustin Publisher, Locust Valley, NY.
15. See Taft, Robert SJ. Annual SS Cyril and Methodius Lecture. Saint Paul's Seminary, Pittsburg. 2001. *Eastern Catholic Theology: Slow Rebirth after a Long and Difficult Gestation*. *Eastern Churches Journal* 8, no. 2 (2001) p 69ff.
16. Narsai Homily XXXII (D) *On the Church and on the Priesthood* in Texts and Studies; contributions to Patristic Literature, edited by J. Armitage Robinson D.D. Volume VIII, *The Liturgical Homiles of Narsai*. Reprint Kraus reprint limited, Liechtenstein, 1967. p 62.
17. Cooke, B., and Macy, Gary. 2005. *Christian Symbol and Ritual*. OUP, Oxford, p 45.
18. Taft, Robert (ed.). 1995. *Liturgy in Byzantium and Beyond*. Ashgate, Variorum, p 47.
19. Taft 1995, p 48.
20. Zizioulas, J.D. 1993. *Being as Communion: Studies in Personhood and the Church*. St Vladimir's Seminary Press, Crestwood, NY, p 17.
21. Schmemann op cit 1974. p 93.
22. Ibid, p 27.
23. Edwards, Denis. 2006. *Ecology at the Heart of Faith*. Orbis Books, Maryknoll, New York. p 69.
24. Schmemann 1966. p 58.
25. Schmemann 1974. p 33.
26. See: The Prayer of the Trisagion in *The Service Book of the Holy Eastern Orthodox Catholic and Apostolic Church*. Tenth Edition, 1997. Antiochian Orthodox Diocese of North America, Englewood, NJ. p 98.
27. See Schmemann op cit p 37ff.
28. Taft, Robert S.J. 2001. *Divine Liturgies-Human Problems in Byzantium, Armenia, Syria and Palestine*. Ashgate, Aldershot. p 29. See also Sozomen. *Hist.Eccles*. VIII,18.7–8.
29. Byzantine Daily Worship. 1969. Alleluia Press, National Publishing Company, Philadelphia, PA. Presanctified Liturgy. p 381.
30. Zizioulas, John. *Preserving God's Creation: Three lectures on Ecology and Theology*. King's Theological Review 12 (1989) pp 1–5.

31. Edwards op cit p 100.
32. Orthodox Service Book p 122.
33. Brockelman, Paul. 1999. *Cosmology and Creation: The Spiritual Significance of Contemporary Cosmology*. OUP, Oxford. p 11.
34. Brockelman 1999, p 157.
35. Spira, Andreas, and Klock, Christoph (eds.). 1981. *The Easter Sermons of Gregory of Nyssa* Patristic Monograph Series, No.9. The Philadelphia Patristic Foundation, Ltd., Philadelphia. p 9.
36. Cyril of Jerusalem. *Lecture XXII. On the Mysteries IV. On the Body and Blood of Christ n 3*. The Catechetical Lectures of St. Cyril of Jerusalem [Kindle Edition] 2013 Edwin Hamilton Gifford (trans) Amazon Media EU.
37. Schmemann, A. 1987. *The Eucharist: Sacrament of the Kingdom*. St Vladimir's Seminary Press NY.
38. Liturgy of St John Chrysostom, p 48.
39. Ibid, p 29.
40. Schmemann 1987, p 225.
41. Ibid, p 226, see also Kenneth Stevenson. 1992. *Worship: Wonderful and Sacred Mystery*. The Pastoral Press, Washington, DC. Chapter 2 *Eucharistic Offering: Does Research Make Any Difference?* pp 23–41.

· 4 ·

PERSPECTIVES ON THE ENVIRONMENT—THE ICON, DOORS OF PERCEPTION

I. The Icon: Door to the Sacred

One area which connects us both to the lived environment and the Kingdom, is that of the detailed and important field of byzantine iconography. There are of course other iconographical traditions in Christianity, the Copts have their own style of icon painting as do the Ethiopian Churches, Latin Christianity shared much of the tradition in its Romanesque style and amongst some scholars there is a debate about a 'tradition' of iconography within the western Catholic Church.[1] In recent years a much greater appreciation for this art form has emerged in more Protestant Christian communities, but to see the development of an art form that is deeply theological and liturgical, we must examine the context of the byzantine tradition because it is there that its role as a medium of the sacred is most clearly established.[2] In this setting of the liturgical and prayerful dimension of byzantine worship, the Icon has always been spoken of as the 'doorway to heaven' or 'doors to the sacred', it is not only an art form but an indispensible expression of Orthodox faith.[3] If we invert the phrase and see the icon not only as a door to the realm of God but as a doorway to human and earthly life, we can also perceive it as the connective point where God, in one of the many aspects of presence, reveals and connects with human beings in their attitude of prayer. There is something deeply theological about this two way process, it has an echo of the Jacob's dream at Bethel (Gen. 28: 10–22) of the angels descending and ascending or his great wrestling match with the divine one at Peniel, a place where he encounters the living God and changes his name to Israel. (Gen.32: 22–32). Most importantly it is also about the transcendent quality of presence revealed in material elements. If the believer, at prayer before the icons, senses that in faith the incarnate God is present to them through this vehicle, the Word not only made flesh but transferred onto material elements, then they also understand that the Icon, like them, has a transcendent quality, revealing through the veiled fashion of image and art the unseen God at the centre and heart of all things, but it also has that quality of hard encounter, spiritually and emotionally wrestling with something almost too powerful to comprehend and yet there with us.

II. The Blessing of Icons: Connectivity

The formula for the blessing of icons is interesting and in its own way can help us lift a particular lid on the intimate connection between environment and prayerful worship, which in the context of the icon is intrinsic and primary. Many people who enjoy icons as an art form and are very appreciative of the message they convey fail to recognize that they are firstly liturgical; belonging to the environmental context of the church building and in the setting of the *iconostasis*. This creates a specific form of environmental encounter that is always part of any type of liturgical action, the locus of the *iconostasis* is a place of supplication, prayer, blessings, intercession and movement. Even in its simplest form of two icons, the Saviour and the *Theotokos*, the icons provide that mediating context between the earth and the heavens which is present in the liturgical action. Secondly the icon is derivative of this liturgical setting because it is prayer that forms the contextual matrix for those in domestic and other settings. It is not for nothing that the house with its prayer corner, and therefore a life of living faith, is known popularly as 'the domestic church'. That is why the icon is blessed in a solemn manner usually within the setting of worship. Here is the prayer of blessing itself:

> Master, our Almighty King, Father of our Lord and Savior Jesus Christ, You gave orders to your servant Moses to sketch a picture of a Cherub in the holy Tent, and from this, we took the custom of sketching icons as a remembrance of those whom

> they represent. Therefore, we pray to You, O Lord our King, to send the grace of your Holy Spirit, together with your angel, on this holy icon so that every prayer which is offered to You through this icon may be accepted by the grace, mercy and compassion of your only-begotten son, our Lord, God and Savior, Jesus Christ, the Lover of Mankind.[4]

If we follow the structure of the prayer several key elements emerge in an evolutionary format. It is essentially a highly Trinitarian formula of prayer, as are so many in the Byzantine tradition. This in itself makes the context that of community, for the Trinity is not only seen and understood as the three persons in one, but also as a community of dialogue and love whose unity is manifest in love itself. If this prayer is addressed to the Father, Master and King, the grace of the Holy Spirit is essential for the blessing and work of the icon in the context of people's ministry of offering prayer through the image, but is also accepted through the love of Christ. Whilst this may seem obvious it does give the icon a specific role in the life and worship of the Church. There is an ambience of holiness that underpins and undergirds the painted icon itself. What this means for our own understanding is quite important for the connection we need to make between the material and the spiritual spheres of life. The great link of course is the materiality of the icon itself. The development and history of this particular art form emerged from several sources but at a time when colour was obtained through natural methods, earth, precious stones, vegetable matter The board itself was made of treated wood, the gesso again a combination of mineral and animal products. All the ingredients, including the person 'writing' the icon were (and are) connected as matter. This in itself helped a theology of icons to grow in which all these connections are seen as part of a whole linked in God. Even the stylistic content is governed by a Canon, rules passed on, a living tradition for the visible word. In turn the one who paints (though the technical term is writes) an icon is sharing in part of God's own creative action. The subject matter becomes a meditation, almost a visual *lectio divina*[5] and if there is a particular person or community commissioning the icon they too are prayed for. Finally of course the icon is blessed before being paced either in church or the home where it in turn becomes the door of perception to the reality beyond itself.

Quite clearly we need to remind ourselves that one of the great characteristics of the east is an insistence on the unity of the Triune God, which is so deeply rooted in the theology of creation, it is evident that this great prayer of blessing is Trinitarian, here we can begin to discern several connections that will eventually link the icon with the environmental and material world of our lives. The prayer is addressed, as are nearly all oriental and eastern Christian liturgical prayers, to the first person of the Trinity, the God and Father of our Lord Jesus Christ, the maker of all things. This directs the iconographer towards the source of all creation and the link that binds us together, it also inserts us into the dynamic love and life of the eternal Trinity but helps the painter understand that they too share in the continual act of creation alongside so many other ways in which God's grace-filled renewal and creative love sustains us .We must note that the action of blessing and dedication is through the grace of the Holy Spirit. One of the gifts of the East is a real love and belief in the power and work of the Spirit. This does not, of course, suggest that forms of western Christianity have any the less understanding, but one of the eastern hallmarks has been that strong defence of the work of the Spirit as an activity of God before time began and continuing through all ages. This was why the issue of the western insertion of the *filioque* in the Nicene Creed became such an important symbol. The Orthodox wanted to safeguard the intrinsic nature of the Spirit as a distinct and separate person of the Trinity so that the post-Pentecost Church is truly understood as that of the Spirit. Then the acceptance and offering of the prayer to the Father is seen as the compassionate and merciful task of the Lord Jesus Christ, described in that very beautiful eastern manner as the 'lover of mankind', this Christ, word incarnate, *Kyrios*, the Messiah is truly at one with and in us making his life our own and his presence continually with us until the end of time. We begin and end in God, this prayer tells us that and it places the whole of our world and the cosmos as part of that life in and with God.

The second aspect of the prayer, which is important for our reflections on the intricacies of the web of divine life, deals with the constant references to the heavenly beings mentioned within its structure. Far from being a simple poetic and symbolic image, which another more rational age can dismiss in ways such as suggestions that this is the genius of earlier people who needed to create some way of explaining God, or a composite of religious traditions synthesized together, we can enter into direct dialogue with the imagery and find in it another reality. The manner in which the prayer links the blessing with the directives found in Exodus 26: 1–14 is not only interesting, but the indication is quite specific in its detail, Moses being ordered by God to sketch the cherubim. The Cherubim are described in both Old and New Testament as having four faces, Ezekiel visions tells us that he perceived them as having the faces of a man, lion, Ox and Eagle (Ezekiel 1.10–14).In the book of Revelation4: 6–8 they are described as four living creatures each having six wings and full of eyes, here they cry out incessantly, 'Holy,Holy, Holy,Lord God Almighty, who was and is and is to come'. These are always associated with the holiness of God but also represent the righteous governance of God and also the exercise of God's judgment. However their images suggest other links as well, their human face connects them to humankind, the animal and bird images wider life connections and the exemplars of strength, height, movement with the attributes of being human, but essentially what is going on is the attempt to express what is almost inexpressible, those creatures are not of human nature or design, but closely connected to the majesty of God. Symbolically they are linked to the attributes of the four Gospel writers and in that sense enhance the powerful theology of the living Word that is God, in Christological terms expressing the presence of Christ himself in the proclamation of the Gospels and in sacramental worship.

III. The Icon: Distraction or Focus?

Is there a danger that these kind of images would be a distraction to believers, take them away from their relationship and focus on God? This is an entirely reasonable question and one which keeps on emerging whenever symbol, sign and image are used to express the being and majesty of the Most High God. After all the underlying prohibition of trying to image God would have been foremost in people's minds. What we can consider is that Ezekiel would have been trying to give a derivative sense of the majesty and power of the Divine presence, it is analogical in context, if the Cherubim are like this, then just imagine what the majesty of the Most High God must be like? The approximation of these beings to God is not accidental, for they stand outside of earthly reckoning beyond the life and experience of mortals, but the capturing of their image as a piece of religious art used to enhance the experience of prayers and worship is very important as the blessing prayer suggests. These heavenly beings are also part of created life in the wider context of cosmos or 'heavenly realms', but the fact that they are seen and experienced in a visionary format places them also as servants of God in life. The value of earthly creation is seen not only in the fact that it is intrinsically good, that these beings are present to it, but also in the fact that it is appointed by God to be the home for living beings which not only include humankind and all that lives and dwells on earth but also the unseen majesty of God hinted at by these visions. The value of the natural creation is revealed in the fact that it is appointed and blessed by God to be the home for so many living beings. The fact that it was also made for God (something which is beautifully expressed in Orthodox iconography), is to recognize our earth as the context for God's incarnation and humankind's deification, and as such, 'the beginning of the actualization of the Kingdom of God'.[6]

The prayer goes on to connect the imaging of the Cherubim with the act of creating, not the primal creation of God's action, but the derivative creation that is tacitly and openly understood in the situation of making, painting and then blessing the icon as sharing in God's action through the incarnation and therefore our natural world. It is also understood as a repetitive act and part of the life of faith, it is the tradition handed on from one generation to the next, but a tradition in the correct theological sense of living tradition, from Moses 'we took the custom' and this lineage continues today and tomorrow until the end of time. In another sense this is an exegesis that tackles the prohibition given by Moses on the worship of idols and graven images. By making Moses the hander on of tradition given by God, theologians like John of Damascus, a great defender of icons against the iconoclasts, were able to allow the use of image in Christian art as a legitimate form not only for representation of Christ as the one who became flesh and as such able to be represented like us, but also for the purposes of worship and prayer. Here is John trying to connecting the threads together:

> How depict the invisible? How picture the inconceivable? How give expression to the limitless, the immeasurable, the invisible? How give a form to immensity? How paint immortality? How localise mystery? It is clear that when you contemplate God, who is a pure spirit, becoming man for your sake, you will be able to clothe Him with the human form. When the Invisible One becomes visible to flesh, you may then draw a likeness of His form. When He who is a pure spirit, without form or limit, immeasurable in the boundlessness of His own nature, existing as God, takes upon Himself the form of a servant in substance and in stature, and a body of flesh, then you may draw His likeness, and show it to anyone willing to contemplate it. Depict His ineffable condescension, His virginal birth, His baptism in the Jordan, His transfiguration on Thabor, His all-powerful sufferings, His death and miracles, the proofs of His Godhead, the deeds which He worked in the flesh through divine power, His saving Cross, His Sepulchre, and resurrection, and ascent into heaven. Give to it all the endurance of engraving and colour.[7]

As John suggests, there are some deep questions to be asked. Is there any way in which human beings can express the ineffable, invisible, inconceivable God? Perhaps we need to step sideways and remind ourselves that imagery is not something distinct and apart from speech and writing. The prayer hints at the method of connection because it frames the blessing in terms of reciprocal action, we make, we offer, you bless and sanctify, both God and the human artist are linked together so that in the setting of prayer a framework for experience is born. John of Damascus points to the bridge, by thinking, by contemplating in prayer the mysteries revealed in scripture, our minds seek to visualize what we hear and experience and that is something God given. I am reminded of a broadcast where a woman explained that she was unable to express what was wrong with her, so drew a picture of her symptoms and took that to the doctor who understood immediately what she meant. What she understood almost instinctively was that an image sometimes explains in clearer and more concise terms the immediate experience.[8]

The theological insistence on the twofold characteristics of biblical reference coupled with the living tradition of the community is one of the hallmarks of eastern Christianity. Just in case we think that the early Fathers might have been stretching a point, especially as Judaism and also later Islam refused to have any truck with portraying God, we have to remember that they saw the Christ event as something that alters human and universal history and changes the focus of the Divine Presence.[9] It is not only an event that has an effect on the 'earth peoples' but is also cosmic in its dimensions. In this sense whatever the meaning was for them of the 'incarnation life passion death and resurrection of Christ' it would also be the hermeneutical lens by which they looked at their world and which we have to understand in order to perceive as they did, but it requires an intellectual and spiritual humility to reach out and experience something familiar in a new way. The monastic tradition, which unites east and west in more than one way, has much to offer in contemplation of this kind of issue. In recent years many monastics have been able to reconnect with the deep and ancient tradition of the Word and the living tradition of their life in prayer both East and West, offer us a new vision of the living word actualized and event-creating in different situations. Father Christian de Chergé OCSO, one of the Cistercian monks of Tibherine who was killed in the Atlas Mountains of Algeria, made famous by the film 'Of Gods and Men', framed this way of doing things in terms of commonality. For De Chergé it was primarily the dialogue with Islam, but intrinsic to his approach is the identity and understanding of the 'common word'. The traditions of Judaism, Christianity and Islam each received the word from God but express that revelation in different ways. It is this ability to transcend division and seek other threads and perhaps 'divine permissions' that enable people to reach out beyond their own confines, not without effort and commonly with misunderstanding, but it has to be done. As the monastic tradition bridges east and west it can dialogue in and through difference, but then find in prayer, worship and meditation part of truth itself and help re-shape our focus. The deep monastic tradition of reflection, but also openness to difference was expressed by De Chergé in this way:

> Human beings come to know mercy if they have the intelligence to discover it in the gift of life, in nature, and in the rhythm of the seasons, but also in the gift of the Bible or the Qur'an…Human beings have the vocation to 'reflect on the merciful presence of their creator'.[10]

This takes us to the heart of why iconography, rooted in eastern patterns of Christianity and, let it be said, fought over in a temporal space not only occupied by iconodules but also Islam, is a theology of word as well as image. In essence each icon, reflecting the reality it represents is also a vehicle of the merciful love and forgiveness of God. We can only be thankful that those ancient fathers and mothers had that instinctive intelligence to understand image not as idol but gift from the creator.

Perhaps this is also one of the delights of discovering the patterns of theology, tradition and prayer from the treasure house of the Eastern Christian experience. It allows us to lift off the shoulders of particular Christian histories our own baggage for a while and listen to those who forged the thought about icons as a powerful theological and spiritual tool. Simeon of Mount Thaumastus understood the iconic form to be something well beyond mere likeness, when praying and worshiping before them he understood this to be not only honouring and worshipping the invisible God, who is not only present but is the 'essence' of all reality.[11] It is this wider and more poetically symbolic encounter with 'the divine other' that helps us grasp something we can intuitively discover in our own lives. What is this encounter? Perhaps we can best represent this in simplistic terms and return to the more complex dimension later. If we can accept television, film, the whole aspect of multimedia technology that allows us to see things we would not normally be able to, such as travel films, documentaries or narrated history and if we accept that in the twentieth and twenty-first century we belong to the photographed generations, and that these allow us to enter into the secrets of others even in an emotional manner, then we perhaps can begin to see how it is that the icon, made by prayer, formed according to a deeply held belief system and located within the context of prayer and worship and not within an art gallery or collection is a locus of deep spiritual encounter.

Meditating on what Augustine of Hippo refers to as the *Verbum Visibile*, the scriptures revealed in the visible word, the believer finds a deep encounter in the presence of the icon. This stress on the word made visible is further emphasized and enhanced in our blessing prayer by the small reference to sending an Angel with the grace of the Holy Spirit to somehow be with the image. Angels are important in the Christian East, not only are they messengers of the word of God, which immediately gives a direct association with iconography and let it not be forgotten a real presence in worship. All this is helpful because it takes away resonances of image worship. In a theological and symbolic manner this small reference tells us that venerating the icon is not idolatry because the icon is not the immediate object, it is the doorway beyond itself or a two way approach to the relationship of the human and material world with God. Melissa Jones writing in the New Catholic Reporter a review of a book of essays on Orthodox perspectives on environmental issues, refers to the distinction between pantheism and panentheism, in other words the problem that occurs when people associate God with the object, God as nature, as sea or whatever, rather than see the essence of God present in things. This is certainly what one of the best loved of liturgical prayers to the Holy Spirit puts it, 'Oh Heavenly King, Comforter, the Spirit of Truth, who art everywhere and fillest all things, Treasury of blessings and giver of life…'.[12] It is the filling all things that takes us away from a false perspective because it avoids any implicit association of the object as God, rather it keeps the dimension of transcendent God and immanent presence in harmony. Jones encapsulates this neatly when she writes:

> …readers learn that Orthodox theology describes God as the transcendent essence, but asserts that every particle (alive or inert) of the material world was created and is sustained by the divine energies of this transcendent, uncreated essence. Orthodox Christian leaders don't spend much time doctrinally arming their flocks against the evils of pantheism because they don't think that creation is easily confused with God the transcendent creator. Here's a metaphor: We can enjoy the warm rays of the sun and know that it sustains life, but a reasonable person would never confuse a warm rock or tree for the sun.[13]

There are of course a few serious issues to sort out with this particular point of view. The identification of God with matter and particularly the incarnation of Christ and its consequences for this type of theology, in particular the understanding of *theosis*, becoming a participant in the life of God so that we become quite literally united with God could, unless it is balanced by a humbler approach, lead to a totally arrogant attitude that humanity is better and above all else because we are those who can share in God. In view of the necessity of recovering a better and more open theology of environmental issues and our concern for the animal world we need these particular groundings in the eastern perspectives on aspects of materiality such as the role and place of the icon, because this will lead us to one of the best corrections for any attempt at domination even in a spiritual sense. This is the great eastern ideal of *kenoticism*, the self emptying of ourselves in love based on Christ's *kenosis*, his emptying out, taking on the form of a servant so that in Him, God reaches the depths of human life even in its most unpleasant aspects. John of Damascus reminds us that this idea of *kenosis* ,emptying out is important in understanding icons, they might be humble images, created and formed by prayerful people but they have a potency that is immense, they are symbols of so much more of a cosmic connection between earth and the Divine Creator and as such he defends their use:

> A small thing is not small when it leads to something great, nor indeed is it a thing of no matter to give up the ancient tradition of all the Church held by our forefathers, whose conduct we should observe, and whose faith we should imitate.[14]

In terms of the blessing prayer we note that Jesus himself is the one who accepts, in another sense is the mediator and in a real sense icons of the Lord Jesus and the feasts are very much a visible gospel, but they cannot be said to be the 'real presence of Christ'. There are several very good reasons for this, firstly one of which is the position of Christ himself, for the Eastern Church he is understood as the incarnate Word, the only begotten son of the Father who as the prayer makes known is the one who accepts and blesses this offering of the images painted for use as a medium of prayer and blessing. Secondly Christ as the icon of the living God, cannot paint or represent himself, for he is the true prototype of all icons, in that sense John of Damascus great phrase reminds us that after his resurrection, until the end of time, he can only be known through the life giving gift of the Holy Spirit. 'I make an image of the God whom I see. I do not worship matter, I worship the God of matter, who became matter for my sake, and deigned to inhabit matter, who worked out my salvation through matter. I will not cease from honouring that matter which works my salvation. I venerate it, though not as God [16]'.

That is the great clue, 'I venerate it, though not as God'. It is in that sense that the blessing prayer safeguards the icon for any connotation of idolatry, its blessing and place is for our use, so we can share in the revelation of the Word communicated to us through beauty, but it is not the true icon of the Lord.

Why? Here is the answer, firstly because the prayer reminds us of the indissoluble nature of the Holy Trinity. Jesus is the Icon of the living God, we by our lives and ministry become living 'icons' of Christ as have all holy women and men, but we do so by the power and action of the Holy Spirit so that we too may know the way to the Father and worship in spirit and truth. In this theological sense the creation of the icon from matter, wood, earth, stone, animal products links it the material world but also through prayer, worship, contemplation enable the believer to lift up created matter into the very life of the Holy Trinity. We become part of the connectivity of creation and God. Professor Veslin Kesich of St Vladimir's Seminary, New York, reminds us of these several links when he writes:

> Christ appears to us in three distinct icons. First, there are his visual icons, which adorn and illumine our churches and homes. The icons of holy men and women are icons of Christ as well, for in their lives they aimed only to 'become Christs in Jesus Christ.' When we venerate the saints, we glorify their Master at the same time, because they were his disciples. By following and committing themselves to him they became 'Christs.'

We also have a third icon, the verbal icon of Christ, which is depicted in the four Gospels.[15]

Part of the quote refers to following and committing actions that mark out the disciples of the Lord, but by extension we can link it to a much wider concept, we have seen how eastern theology places the icon in a position of encounter, where the material and created object itself points to the reality behind image, that of life and God but is also a locus of that meeting. If we accept this then all that we can see or experience within the created order of things has that potential and ability to become part of the encounter with Christ. In other words all that lives and all that is part of our natural and created world becomes part of the greater design each part linked to the other in creation itself as an icon of God. In a sense this concept removes a problem from our midst. There is a tendency to locate holiness in a few objects, invest them with a sacrality that places them apart from other forms of material life, the icon helps us modify that standpoint. It is important for us all to recognize that William Blake's poetic imagery of the 'World in a grain of sand', found in his 'Auguries of Innocence':

> To see a World in a Grain of Sand
> And a Heaven in a Wild Flower,
> Hold Infinity in the palm of your hand
> And Eternity in an hour,[16]

is not just a Christian concept of implicit holiness to be mystically encountered in all things, but more specifically what the Christian accepts through the Incarnation of the Lord and the gifting of the Holy Spirit as a continual locus of presence realised amongst us. In this sense there is an urgent need to be more pro-active in proclaiming that insight. Part of that mission given to the disciples of Jesus is to follow his presence as it is revealed by the Spirit in our whole world and commit ourselves to showing others that our world needs careful handling and deep respect as it is a medium of this sacred presence.

IV. The Icon: Door of Perception

What does the technique and style of iconography offer to our thoughts on environmental issues? I have hinted that the key is found in the phrase 'door of perception'. We noted that the blessing of an icon invests an extraordinary theology to the symbolism and meaning of the object, but it is also important that we understand why the process, the various materials used, the structural design and the particular subjects that are allowed (and by default not allowed) combine to form a very powerful theology of matter. In this sense understanding the icon can help transform and renew our own attitudes towards world and matter as well as all living creatures. To begin with this art form is rooted in antiquity and as such has preserved the usage of natural pigments, glues, varnish and such like, whilst we can understand that there is a long tradition of this form of gesso and tempera painting in art, the fact that the icon retained this format when newer methods and materials had appeared suggests a conscious and deliberate policy. There was, of course, a period when alien elements found their way into the art form, but this, like theology, was subject to strict scrutiny by the Church herself and anything that seemed to deviate from the authentic canon including subject matter was understood not to be part of the living tradition in all senses of the word.

But what is the 'authentic tradition of icon writing'? Is it not simply a human art form that has taken on the patina of holiness by association and symbolism with the Christian Church developing as time went on and doctrines became more codified into a complex and highly symbolic set of imagery? The historical roots of eastern icon painting goes far back into cultures such as those of the Babylonian, Persian and ancient Egyptians. There is a direct connection with tomb portraits found on mummies at Fayum and elsewhere, but whatever the origins and connections it still remains and indisputable fact that in Christian hands the icon became a uniquely spiritual art form and can never be disassociated from its context of liturgy and prayer.[17] In terms of the connectivity with our concerns and hopes for the environment of the world and cosmos, these supremely earthly pictures connect us to the very dust of the cosmos but also to the transfiguration of life in Christ.

Many books have been written about the tradition of authentic icon writing, in the Russian tradition Vladimir Lossky was one of the most influential theologians to encourage wider understanding of the role and place of the icon in the Church.[18] In his book 'Traditions and Traditions' he writes:

> Just as much as dogmatic definitions, it has been possible for the icons of Christ to be compared to Holy Scriptures, to receive the same veneration, since iconography sets forth in colors what the word announces in written letters... Icons impinge on our consciousness by means of the outer senses, presenting to us the same suprasensible reality in 'esthetic' expressions (in the proper sense of the word *aisthetikos*—that which can be perceived by the senses). But the intelligible element does not remain foreign to iconography: in looking at an icon one discovers in it a 'logical' structure, a dogmatic content which has determined its composition. This does not mean that icons are a kind of hieroglyph or sacred rebus, translating dogmas into a language of conventional signs. If the intelligibility which penetrates these sensible images is identical with that of the dogmas of the Church, it is that the two 'traditions'—dogmatic and iconographic—coincide in so far as they express, each by its proper means, the same revealed reality.[19]

Whilst this links dogma, doctrine and icon it has a very useful point for our reflections on the environmental issue. Words are just one part of the richness of communication between human beings, but because there are many ways of experiencing human life, especially when we get old or find that we are working within limitations such as sickness or injury, many other way of connection can be found such as the use of visual images, a variety of symbols, gestures that speak without word or enhance it, smell, hearing, touch, in short the whole orbit of the senses. It is because we inhabit a world of sensory experience that the byzantine tradition of iconography recognizes this as an essential and important gateway to another realm of experience. The 'announcing' in colours, as Lossky determines it, forms part of the essential communication of the icon as 'word'. Just as the Holy Scriptures are read and broken in the homily so that people can be fed by them so it is with the icons use of colour and form.[20] Because they are made from natural elements, pigments which are then ground up and refined, bound together with water and egg yolk they relate to the intimate connection of human activity and nature, but because after this process they are then used they take on a homiletic dimension, 'breaking' the word of God through the image perceived and meditated upon. If the truth be told the colour scheme of the icon is important in prompting a reaction in the person praying or looking at the image. The settings are not meant to be realistic, in fact they could be described as 'No' theatre, the bare essentials pointing to the key character, the gesture of the hands, the draperies, the blocks of different colours all speak of different meanings.

This tradition of each colour speaking to us is ancient, but rather curiously the impact of the colour sequencing gives an almost aniconic quality to the subject matter, a luminosity beyond form that is sometimes glimpsed in the true liturgical setting, when the lamp light catches the gold leaf and fills the icon with a brilliance or found in the gaze of the Christ whose eyes look right through the viewer. This activity of encounter and transcendence is very reminiscent of what the great french stained glass artist, Gabriel Loire (1904–1996) spoke of the power of colour when he commented on the design of his windows for Saint Mary's University Chapel Twickenham. Having been commissioned to design windows based on the Catholic Mysteries of the Rosary in his specialized *Dalle* glass, he spoke of his desire to let the glass colours speak for themselves.[21] In other words to become part of the experience of the person praying or looking in an encounter which at its heart had the experience of light, for him the windows were simply intermediaries between Gods light and our own life.[22] Because Loire's manner of glass making utilized medieval techniques, fusing the colour into the glass, he was able to create multiple hues and so simply used blocks of blue, red, white gold and brown glass formed into an abstract design, which coupled with the use of accessible symbols of the Virgin Mary and of Christ placed in such a way as to hint at the mystery portrayed.[23] Though he was not a byzantine iconographer his insight is precisely that of the icon writer, whose use of paint and form allows the transcendent meanings to be found in an experience of light itself, refracted and reflected, thus enlarging the experience of the person in contact with the icon.

> The Icon's background above all reveals the 'Uncreated Light'…The background never merges with the image. It is always separate, even if it is not just gold or silver but just a colour. At the same time, however, the background and the image form an organic whole. Our Lord is both infinitely higher than mankind and wishes to be united with us. Hymnography calls Christ the 'Sun of Truth.' This sun lights the whole world, but the saints transfigured by it become light-bearing themselves and shine like heavenly bodies in the firmament.[24]

The theology of light (*hesychasm*) is one of the most influential aspects of eastern spirituality which flowered under the teaching of Saint Gregory Palmas and the period of fourteenth and fifteenth century byzantine iconography which focuses in a particular way on the icon of the Transfiguration. In an examination of iconographical tradition and the insights we may gain from it for our 'gaze' on the environmental world the importance of colour as part of that transfigured brilliance cannot be underestimated. The meaning given to them is highly important and so their use (much as Loire discovered) follows a particular canon so that we can connect both our world and that of the divine realm. David Morgan in his book, 'The Sacred Gaze', tries to create a definition of religious visual culture. Part of his thesis is to examine the concept of 'gaze' which is, as we can understand is very much part of the veneration of icons, we gaze on them, but as Morgan points out this carries with it a whole array of meanings that we bring to the act. He talks about the wide variety of visual practices which engage the person in looking at objects especially in a religious setting, but reminds us that a certain amount of caution is necessary; seeing is about a visual culture which needs to be understood in its context. Gazing on the icon for instance involves artifacts, tools, assumptions, promptings and a whole realm of configurations within a liturgical and theological matrix. There are rules in the approach to iconography as a spiritual vehicle in which the visual retains parity with the written and verbal forms of communication. This is one of the theological gifts of the East: 'Christian orthodoxy had defined itself in creeds but had resisted allowing verbal utterance to eliminate visual piety. Images belonged to the profession of faith. Seeing, no less than saying, was a fundamental, even liturgical, aspect of believing'.[25]

V. The Icon as a Moral Compass

One aspect of the Icon that we can distinguish in terms of its influence on the believer is its capacity to proclaim particular moral truths. In a lecture on Icons given at St Vladimir's Seminary in New York in 2011, Metropolitan Hilarion Alfeyev of the Russian Orthodox Church spoke about the various ways that the Icon can affect touch upon the lives of eastern Christians, and by extension others. In particular he points out the aspect of morality which iconography represents in its theological context. For instance in an icon such as the Dormition of the Virgin Mary each part plays a role in opening a particular theological perspective on the role of death. Alfeyev points out that the way the Mother of God lies peacefully stretched out on her deathbed, surrounded by the apostles and angels who venerate her body, as we too are venerated at our funeral service, above them is the Risen Christ receiving the Virgin as an infant in his arms. Thus does the icon avoid any sense of dualism, the dead Mary is somehow being transfigured in death from the mortal life we lead to that immortal and new life promised us by Christ. Some interpreters suggest that the infant is the soul, but that would detract from one of the fundamental beliefs found in the creed, that the Christian believes in the 'resurrection of the body'. In one of his general audiences Pope Francis reflected on the phrase in the Nicene Creed about the resurrection of the body and in his own catechetically direct manner reminded people how much of our contemporary society fears death: 'I have always been struck by the question: why do children suffer?, why do children die?,' he said. 'If it is understood as the end of everything, death frightens us, it terrifies us, it becomes a threat that shatters every dream, every prospect, that breaks every relationship and blocks every path.'[26] But said the Pope the Christian should not fear death, if our lives have been a 'journey with the Lord' then our end will be 'to contemplate face-to-face that wonderful countenance of the Lord, to see him as he is, beautiful, full of light, full of love, full of tenderness. We go towards this end: to see the Lord.'[27] The catchy phrase he used to sum up his reflection is almost a liturgical acclamation: 'One who practices mercy does not fear death', for it captures the essence of what those icons of the Resurrection and Dormition are conveying in the *verbum visibile*, that in Christ mercy and justice met, taking on human life the Lord brings it to completion and fulfillment in his resurrected presence, when all will be made new. His command to love is borne out in discipleship especially that of Love of God, neighbour (which includes creation) and our very selves.

Essentially, every icon bears a powerful moral charge. The icon reminds modern man that, apart from the world in which he lives, there is also the other world; apart from the values preached by irreligious humanism, there are other spiritual values; apart from the ethical norms established by secular society, there are other standards and norms.[28] Patriarch Bartholomew invariably relates the environment to a familiar aspect of Orthodox spirituality, namely to the icons that decorate Orthodox churches. Symbols are important in Orthodox thought, worship and life. Creation itself is likened to an icon, just as the human person is created 'in the image and likeness of

God' (Gen. 1.26 and Col. 1.15). The Patriarch invites people to contemplate the Creator God through the icon of the created world (Col. 1.16–18). Creation is a visible and tangible revelation of the presence of the Word of God.

In his view humanity is called to wonder at creation not in any romantic sense but in that engagement of proper awe, but on the other hand we are not called to worship creation, otherwise the world of nature is 'reduced' from the level of icon where it mediates the presence of the holy, being in that sense of encounter, part of the dynamic of the relationship we have with each other and God to become a narrow and confining idol which transfers worship of the creator to the created. Alefeyev remineds us that: The icon reflects the eschatological, apokatastatic, redeemed and deified state of nature.[29] In the same vein, Patriarch Bartholomew refers to the human beings as endowed by God to serve as 'priests', underlining that personal responsibility for the physical world and the slightest action of even the feeblest among us can change the world for the better: Whenever we narrow religious life to our own concerns, then we overlook the prophetic calling of the Church to implore God and invoke the Divine Spirit for the renewal of the whole polluted cosmos. For, the entire world is the space within which this transformation is enacted. When we are transformed by divine grace, then we discern the injustice in which we are participants; but then we will also labor to share the resources of our planet; then, we realize that eco-justice is paramount—not simply for a better life, but for our very survival.[30]

VI. The Icon as Part of the Name of God

In recent years there has been an unhappy resurgence of fundamentalist Islam in the regions of ancient Christianity. The rise of IS and the barbarities inflicted on populations in Iraq and Syria bear witness to a new and yet ancient struggle, that of one ideology claiming rights over others. In all this horror one clear focus has been the misuse of the name of God. To claim, as people have done and still do, that it is God's will to kill, maim, persecute, destroy and murder innocent people because they do not share another's vision, belief or ideology, is nothing short of blasphemous. This is an evil unleashed on the world and it distorts the image of a Creator concerned and at work with all of life Here is a God of destruction, punishment and law.

Notes

1. See the Orthodox Arts Journal website http://www.orthodoxartsjournal.org/topics/iconography/. For instance see Aidan Hart's article: *Can Statuary Act as Icon?* January 30th 2013.
2. Eveseyva, L. 2005. *A History of Icon Painting*. Grand Holdings Publishers, Moscow.
3. In considering this tradition 'orthodox' also includes the eastern catholic churches who follow the byzantine rite and who utilise icons in exactly the same way.
   See also Yazykova, Irina. 2005. *The Theological Principles of the Icon and Iconography* in Cook, K. (trans). *A History of Icon Painting*. Grand-Holding Publishers, Moscow. pp 11–23.
4. Melkite Formula for the Blessing of Icons taken from the Byzantine Melkite *Euchalogion*.
5. *Lectio divina* is the meditative and prayerful reading of scripture. It is a method of prayer that comes from early Christianity and gained popularity in the monastic tradition. The Rule of Saint Benedict makes it an intrinsic part of the monastic day.
6. Orthodox Perspectives on Creation 1.7.
7. Saint John of Damascus. *Apologia Against Those Who Decry Holy Images*. Part 1. *From St. John Damascene. On Holy Images*. Translated by Mary H. Allies. Thomas Baker, London. 1898. Scanned by Gabriel Caswell and Dr. Stephen J. Shoemaker. Prepared for HTML by Dr. Stephen J. Shoemaker.
8. BBC Radio Four. Womans Hour. Broadcast November 1st 2013.
9. For a succinct theological comment on this in reference to Icons see: Vrame, A.C. 1999. *The Educating Icon*. Holy Cross Orthodox Press, Brookline, MA. Chapter Two. The Theological Rationale of the Icons. pp 21–57.
10. Salenson, C. 2009. *Christian de Chergé: A Theology of Hope*. Liturgical Press, Collegeville, MN. p 45 also quoting De Chergé, *L'invincible espérance*, 90.
11. *With the material picture before our eyes we see the invisible God through the visible representation, and glorify Him as if present, not as a God without reality, but as God who is the essence of being*. Simeon of Thaumastus found in John of Damascus Part III. 141.
12. Byzantine prayer to the Holy Spirit used by both Orthodox and Eastern Catholics who follow the Liturgy of Saint John Chrysostom.
13. Melissa Jones book review, *A Divinely Written Icon: Orthodox Theology Can Lead to New Thinking on Creation*, in National Catholic Reporter. September 18th 2013. A review of *Toward An Ecology of Transfiguration: Orthodox Christian Perspectives on Environment, Nature, and Creation*, edited by John Chryssavgis and Bruce V. Foltz, published by Fordham University Press.
14. Saint John of Damascus. *Apologia against Those Who Decry Holy Images*. Part 1.
15. Kesich, Veslin. *The Verbal Icon of Christ*, in Jacobs Well, Newspaper of the Diocese of New York and New Jersey. Orthodox Church in America, Winter 1996.
16. Blake, William. Auguries of Innocence.
17. There are some excellent examples of these mummy portraits in the British Museum London.
18. Ouspensky, L., and Lossky, V. 1982.The *Meaning of Icons*. St Vladimir Seminary Press, New York.

19. Lossky, Vladimir. 1974. Tradition and Traditions. Chapter 8 of In the Image and Likeness of God. St Vladimir's Seminary Press, Crestwood, NY. p 167.
20. See the connection between the evolution of iconography and book illustration in Lassus, J. 1966.*The Early Christian and Byzantine World.* Paul Hamlyn, London. p 81.
21. Dalle glass.
22. Loire Glass TWICKENHAM
    Saint Mary's University College Chapel
    Strawberry Hill, Waldegrave Road
    A.E. Richardson et P. Holland arch
    Dalle 1963 à 1965 15 baies 118 m^2.
23. This is taken from an unpublished manuscript at St Mary's University College, Twickenham, UK and is based on the design brief given and worked on in discussion with Gabriel Loire. The stained glass there has one of his finest windows, the great west end Coronation of the Virgin. Loire wanted to let the windows speak for themselves without any imposition of his own. His understanding of colour and what it represents, humanity, divinity etc. is very closely linked to the understanding of colour in the Icon.
24. Yazykova, I. 2007. *The Theological principles of the Icon and Iconography*. p 16. In Cook, K. (trans). A *History of Icon Painting.* Grand-Holding Publishers, Moscow. pp 11–23.
25. Morgan, D. 2005. *The Sacred Gaze: Religious Visual Culture in Theory and Practice*. University of California Press, Los Angeles. p 11.
26. Pope Francis. November 27th 2013. Weekly General Audience. Vatican Rome. See also. La Stampa.it. Vatican Insider. November 27th 2013. www.vaticaninsider.com. The Pope's Speeches.
27. Pope Francis November 27th 2013. See also National Catholic Register: http://www.ncregister.com/daily-news/pope-francis-cautions-against-trivialized-vision-of-death-without-hope#ixzz2m1zGCrci.
28. *Metropolitan HILARION (Alfeyev)*. Theology of Icon in the Orthodox Church. *Lecture at St. Vladimir's Seminary, 5 February 2011*. Department for the External Church Relations of the Russian Orthodox Church, https://mospat.ru/en/2011/02/06/news35783/.
29. *Ibid*.
30. Chryssavgis, John. 2008. *Cosmic Grace – Humble Prayer: The Ecological Vision of the Green Patriarch Bartholomew I.* Eerdmans Publishing Company, Grand Rapids, MI (2nd edition, forthcoming). See more of Patriarch Bartholomew's work at: http://www.patriarchate.org/documents/ecumenical-patriarch-bartholomew-insights-into-an-orthodox-christian-worldview#sthash.WZQIbe4z.dpuf.

· 5 ·

MONASTIC SIGNPOSTS

The tradition of christian monastic spirituality and life and the variety of religious experience that it represents can at first be a bewildering subject for anybody trying to discern just what this phenomenon of Monasticism is about! The spiritual path that it represents spans nearly 2,000 years of christian history, combining both eremitical and community dwelling monastics, both men and women, within the traditions of eastern and western Christianity. There have been periods when monasticism has flourished and really enriched church and world, but there have also been times of decadence and collapse. Whilst we are examining the christian context and the insights of monasticism for our present world and its problems with environmental and global climate change, this way of life has a wider ecological and ecumenical out reach for it builds bridges across religious divides to connect with other religious traditions such as Hinduism and Buddhism. From its origins the example and witness of the first monks was a powerful countercultural voice against an establishment Church and the corrupt Imperial world, whilst this was reactionary and manifest itself in flight from urban centres to desert solitude (soon broken because of the popularity of this way of life) It also had and still has a prophetic voice for contemporary culture because it takes us out of our comfort zones to look again and what and who we really are and in the case of our environmental and ecological issues it challenges us to see again our relationship not only to God but to creation and life! Pope Francis in his encyclical *Laudato Si* points out the role monasticism might play in dealing with our global problems:

> We can also look to the great tradition of monasticism. Originally, it was a kind of flight from the world, an escape from the decadence of the cities. The monks sought the desert, convinced that it was the best place for encountering the presence of God. Later, Saint Benedict of Norcia proposed that his monks live in community, combining prayer and spiritual reading with manual labour (*ora et labora*). Seeing manual labour as spiritually meaningful proved revolutionary. Personal growth and sanctification came to be sought in the interplay of recollection and work. This way of experiencing work makes us more protective and respectful of the environment; it imbues our relationship to the world with a healthy sobriety.[1]

John Chrysostom, though not a monastic in the sense of belonging to a community, nevertheless engaged with the tradition and for a while lived as a hermit, and offers us this insight, which is completely rooted in the Jesus tradition of discipleship. It touches on the complex issue of the human persons place in the world and where in that space the human fits, above it, in charge of it, or as we shall discover in relation to it! Chrysostom is talking about ownership (and from that context the derivative issue of biblical stewardship): 'Tell me, then, whence art thou rich? From whom didst thou receive it, and from whom he who transmitted it to thee? But you must go back and find the original owner!'[2] Whilst this might at first appear contextually vague, if we push the analogy of seeking for meaning, truth and justice, and as the monk does for God, then this is very much part of that 'long search' for meaning in life! The value of monastic insights is that they challenge us to look again at the status quo whilst seeking new ways of expression. To find the original owner of our world and its life is the deep search of the monastic, but it is never one truly taken alone because the monk shares this search with others done for the life of the world! At the heart of this is the spiritual tradition of discernment which is bound up with the vow of obedience, not the slavish autocratic commands of superior to subject, but the deep rooted listening to each other and God to discern what is right and necessary at this moment in time. John Cassian, quoting Abba Anthony of Egypt writes of it in this manner:

> It is discernment which in Scripture is described as the eye and the lamp of the body. This is what the savior says, 'Your eye is the light of your body, and if your eye is sound then there is light in your whole body. But if your eye is diseased then your entire body will be in darkness' (Mt 6: 22–23) This eye sees through all the thougl and actions of a man, examining and illuminating everything which we must do. An

> if it is not sound in a man, that is, if it is not fortified by good judgment and by well-founded knowledge, if it is deluded by error and presumption, this makes for darkness in our entire body.... Hence it is very clear that no virtue can come to full term or can endure without the grace of discernment...For discernment is the mother, the guardian, and the guide of all the virtues.[3]

It is by recapturing other alternative insights like this, that the monastic tradition helps open out the conscience of humanity. Whilst the West seems tired and religious orders struggle against all kinds of other value structures, the eastern tradition of *monos* can be a force of great vitality. This, I believe, is partly due to the positive enrichment of monastic thought and prayers through the interaction of various traditions that has taken place in the eastern setting. Whatever one likes to think of organised religion, the Christian Church has always shared its theology outwards, to the world, not only as thought but also as *praxis*, moreover it has always embraced a wider vision of the Kingdom of God interacting and being part of life. Because of its God-centeredness, particularly in its dynamic understanding of the Trinity and the place of Christ and the working of the Spirit in life, monasticism in its origins and ancestry has always been deeply involved with the created world down the ages and calls the Church back to that connectivity. The monk's calling cannot be truly separated from world, for the monk is a human and lives amongst others needing food shelter, nourishment and companionship to survive. Discernment is part of this legacy more necessary than ever in the complexities of our twenty-first century global problems. Whilst the thought, life and witness of eastern monks and nuns have been rich, and their vibrancy in responding to new horizons and challenges has brought about many different insights, the reality is often complicated by the differences existing between various forms of life in the different Churches and their particular spiritual viewpoints.

We often talk about Monasticism in the Church and then focus closely on the West particularly about monastic families inspired by Benedict of Nursia's Rule or, from one particular form of Eastern Orthodoxy, focus on the phenomenon of 'Athonite' monasticism and theology, and of course their various derivative offspring, but ignore the many other riches known and unknown in the great tradition of the East. Fortunately there has been in recent times a reawakening of interest in the desert tradition and other forms too, all this helps ground our research and helps inform a richer and deeper general picture of the monk. But there is a wider implication for the future of religious dialogue and impact on the life of the world; particularly those issues concerning environment and global change, in the widest sense it can also be an intensely ecumenical movement. In origin Christian Monasticism belongs to the East but developed in particular ways in the West, nevertheless both are part of the same family and build bridges between what might seem disparate and oppositional divides, it is because of this that I have partly merged both traditions in this section of the book.

The movement has particular spiritual insights for environmental issues, but human nature being what it is, as the late Downside Abbey artist-monk, Dom Hubert Van Zeller stressed in his famous 'Brother Choleric' cartoons: 'there is a tendency among monks to…' and I would include 'and people today' to think in particular ways and to follow or react to the promptings of the ecclesial and social situations in which we find ourselves sometimes without deeper reflection.[4] We are all creatures of our time. Recent '*aggiornamento*' in the Catholic Church forced religious communities to re-examine their origins and the charism of their particular founder and to renew their institutions with an eye on the needs of contemporary society and this had an effect on many protestant and some eastern religious communities. Yet this is not enough, a quiet still voice calls us out to move beyond even this, to return to the true source, hence the search for forms and evolutions of the monastic ideal in settings different from the accepted norm, including more lay involvement and counter cultural life styles. In an age when we are faced with huge questions about sustainability and the future of our to planet, Monasticism at its roots can still inform challenge and present alternative ways of thinking living and being. Here is a small taste of such insight, in one of his homilies on the letter of Timothy, John Chrysostom preached something that resonates with those of us concerned about distribution of goods and the aggregation of power but because it is part of the 'tradition' is an important voice which we will find helpful: 'Goods are named from use, not from lordship, and are not our own, and possessions are not a property but a loan. There is a sensible proverb, "O field how many men's hast thou been, and how many men's wilt thou be?" This we should say to all our houses and our goods'.[5] John Cassian also holds us to account, his comments echo those found in Chrysostom and others, but intuitively Cassian understands that not only renunciation of goods and wealth is desirable, but stresses the use to which these goods MUST also be put, for the monk does not simply renounce, but actively embraces the equality of distribution according to need, a principle that is very much at the heart of environmental issues:

> Now when we abandon these visible riches of the world, it is strange goods and not our own that we are leaving. And this is so even if we boast that we acquired them through our own efforts or that they were passed on to us as an inheritance. I say nothing is ours except what is in our hearts, what belongs to our souls, what cannot be taken away by anyone.
>
> Christ speaks in criticism of those who cling to the visible riches of the world as though these were their own and who refuse to give a share to the needy: 'If you cannot be trusted with what is not yours, who will give you what is your very own?' (Lk 16: 12)[6]

The insight here is not simply that of renunciation and giving of goods to the needy, but of what riches actually mean. Cassian, building up on desert teaching, reminds us that there are three types of riches, good which are connected to virtue and merit acquired by a life focused on following God; bad which drag us down, consume us and lead to evil actions and neutral riches which we can use in one way or another depending on who we are. Nevertheless the tradition places the greatest importance on renunciation for the Kingdom as something positive, we give up and give to, 'sell all you have and give it to the poor' (Mt 19: 21) By an inner transformation we start an outer change.

'The apostle had something to say about this appearance of renunciation: "If I give all my goods to feed the poor and hand over my body to be burned and have no love, it is of no benefit to me" (1 Cor 13: 3). The blessed apostle would not have said this if he had not foreseen in spirit what was to come, if he had not known that there would be some who would give al they possessed to feed the poor and would, nevertheless, fail to achieve gospel perfection and the high summit of love, and all because they continued to hold on to the sinful practices of their past, their unrestrained ways, with pride and impatience mastering their hearts, with no thought of how to cure them. And because of this they could never reach the unfailing love of God'.[7]

The question that these ancient texts and sayings puts before our problematic world is a simple one, direct like the word of the abba or amma to their disciple, 'what is our real motivation?' It is also the great call to *metanoia*, to a radical conversion of life for the good of the world and the future, it is to see with different eyes a different vision of life, as saying attributed to Pachomius suggests: 'If you see a man pure and humble, that is a great vision. For what is greater than such a vision, to see the invisible God in a visible man?'.[8]

I. The Search for God

This call by the Holy One to step outside the present pattern is nothing new, for instance in the Hebrew scriptures, Elijah called by the Lord from his despondency is made to encounter the Divine Presence anew, it is only after a few false starts that the discernment of purpose and encounter with God is achieved and it certainly isn't to return to the life of the early Patriarchs, but to move onwards into new paths with the true source of all hope and life, that is with God. The passage about this encounter on Mount Horeb is a classical text for the pilgrim '*monos*' search for God: 'Go and stand on the mountain before the Lord, for the Lord is about to pass by.' Now there was a great wind, so strong that it was splitting mountains and breaking rocks in pieces before the Lord, but the Lord was not in the wind; and after the wind an earthquake, but the Lord was not in the earthquake; and after the earthquake a fire but the Lord was not in the fire; and after the fire a sound of sheer silence. When Elijah heard this he wrapped his face in his mantle and stood at the entrance to the cave. Then there came a voice to him that said, 'what are you doing here Elijah?' (1 Kings 19: 11–14) and after a dialogue a new command and vocation is given "Then the Lord said to him, 'Go, return on your way'" (I Kings 19: 15). The same is true of the disciples of Jesus, exemplified in a particular way by the story of the rich young man which has always been a seminal monastic text in the life of St Anthony: 'If you wish to be perfect, go, sell your possessions and give your money to the poor, and you will have treasure in heaven, then come, follow me'. (Matthew 19: 16–22). All disciples of Jesus are asked to leave all and follow him but not into other people's 'tried and tested' ways of discipleship, rather into a new and total commitment with him to proclaim the Kingdom in their situation.

The desert monks knew this well. A story about Abba Agathon shows how the ascesis of the desert opens out deep horizons, the connectivity of the monk with who he is, and his place is obvious:

> It was said concerning Abba Agathon that some monks came to find him having heard tell of his great discernment. Wanting to see if he would lose his temper they said to him 'Aren't you that Agathon who is said to be a fornicator and a proud man?' 'Yes, it is very true,' he answered. They resumed, 'Aren't you that Agathon who is always talking nonsense?' 'I am.' Again they said 'Aren't you Agathon the heretic?' But at that he replied 'I am not a heretic.' So they asked him, 'Tell us why you accepted everything we cast at you, but repudiated this last insult.' He replied 'The first accusa-

> tions I take to myself, for that is good for my soul. But heresy is separation from God. Now I have no wish to be separated from God.' At this saying they were astonished at his discernment and returned, edified.[9]

This is an expression of true humility which can be translated (with a degree of care) into our engagement with the environment. One of the problems that seems to be endemic in western culture is a 'tendency' to lay blame on everything apart from ourselves, this can also have another side effect that of not repudiating what is definatly wrong. For Agathon the line in the sand was the hersey charge because as he saw it, that was one thing he was not! The monastic search for God can help us create a new psychology of metanoia, conversion of life into truth because it calls one to seek truth. It is this vocation to risk and trust that continually appears in the story of God's friends. I found an echo in the story of the Cistercian monks of Atlas, struggling to live the Cistercian ideal in a different way. Writing about the Prior, Fr Christian, reflecting on a meeting of superiors, Jean Maynard gives this insight, 'The gathering seemed to him to be dominated by the Abbots of large and important monasteries from the rich developed world—a whole monastic scene, complete with grandiose ceremonial which he saw as of the past rather than the future. Himself he saw as representing the voices of Vatican II and the Third World…A key point on his personal agenda was to obtain a small but significant amendment in the Order's Constitution. Trappists were officially supposed to offer monastic hospitality to 'those who share our faith'; he wanted this changed to 'people of goodwill' as a way of opening out the encounter to those who seek truth with us, no matter who they are or where they come from'.[10] Why have I started to think like this? Partly because monastic life is so topical, but partly because we seem destined to keep returning to this type of nostalgic reflection unless we let the living Spirit empower men and women to take up the call to seek God anew and return not to an original rule or the founders vision, but to use these as ways of tracking down God, who in Francis Thompson's words found in his poem, 'The Hound of Heaven', is seeking us:

Halts by me that footfall: Is my gloom, after all, Shade of His hand, outstretched caressingly? 'Ah, fondest, blindest, weakest, I am He Whom thou seekest! Thou dravest love from thee, who dravest Me.'[11] These insights are not external to the human condition, they are part of the fabric of meaning. Necessary for us if we as humans are going to do anything about our planet or redress the terrible wrongs we have done to earth and its creatures.

II. Stepping Outside the Known Boundaries

A *monos* is a person who is trying to live a certain way of being before God, that's almost the bottom line, unfortunately we then get caught up in analysing just what that can be and perhaps in so doing lose sight of what it is all about. The Eastern tradition in the Church can be very helpful here, for this story takes us away from the structured organisation of the west into a more muddled world, a world where strange experiments were tried, some succeeding and some destined to failure. How many of us would see 'fasting' as a totally integral part of being *monos*? Yes I know a number would, but in the language and discourse of classical monasticism it doesn't feature as a first order rule, it is not the total essence—yet it was a singular part of Christian life in the ascetical discipline of striving to be perfect, and dare one say it there were people who made it a total package, in other words made it part of their way! You have only to glimpse through the great tradition of the desert mothers and fathers to see oddities, quirky (to us) ways of living, *dendrites* in trees, those who crawled to be like and with the animals, the procession of holy idiots and fools living a really contradictory form of life and of course those who managed it with a partner (as is found in part of a Celtic tradition) as well as the vast array of those for whom the witness and martyrdom of the monastic way is the monotony of daily regularity and balance![12] These figures are all part of the ancestral DNA of monasticism, but except for the last group, few would advocate an unbridled emulation of them today. It is noticeable in our media-conscious and visually packaged society how terms and images of antiquity market a sense of reverence but also of the exotic when dealing with heritage items, especially in the case of monks and monasteries. It also because of popular perception canonises a particular vision of monastic life.

This is a displaced perception of our cultural and religious antiquity, one that mainly comes from the second millennium of the Christian tradition, not the first, and it can, if left to dominate, lead to a narrowness of vision. I can give an example from my own experience, in a newsletter of 2009 an Anglican Benedictine Abbess wrote about her diminishing communities activities but also of their own vocation, she was disturbed, I think, by reading or perhaps encountering 'new forms of monasticism', mainly the radical new monasticism. In her letter she praised the community fidelity to the 'traditional' form of monastic life, seeking judiciously to distance it from what she had heard of and perhaps met as 'new monasticism'.[13] There is nothing wrong in that, except for inexact terminology, she saw her community (habit-veil-divine office-common life with all the Benedictine trappings) as 'the' ancient tradition, whereas another reading of the DNA, another trawl down the ancestral road would have a said that perhaps these new forms were in fact a throwback to something more ancient and important, and that what she understood as valuable was just one part of a bigger picture, a contribution of the Middle Ages to Western Christendom responding to change and transformation at that time, what some call (again with a trace of irony) a 'classical' model. This is part of a perennial problem when any one model dominates and becomes part of structure crowding out other authentic forms to the periphery. The antidote is to remind ourselves that life is not a static event but in Blessed John Henry Newman's dictum a changing world; 'To live is to change, and to be perfect is to have changed often'[14] and the evolution of particular forms of life have validity in terms of what they are, do and represent, but we cannot claim one authentic strand as a singularly unique form unless it is of course the Gospel way.

III. The Different Experiences of the Tradition

This brings me to examine just what might be a way of convergence between the desire of many people to use the monastic experience and experiment with it as a 'way of being' in new forms of life but also seek a foundation in that DNA based in the long and broad history of Christian tradition both east and west. In the Syrian tradition the word for monk (*ihidaya*) hints at something essentially pilgrim-like, we often define the word monk as one seeking God alone, through a disciplined life often with elements of practical asceticism aimed at a general sense of purification (*ascesis*) and a trust in the promptings of Holy Spirit, but this Syrian hint widens the meaning and points to a deep yearning for some kind of 'return', a return to our original unity with God, what Marilyn Dunn describes as 'oneness and unity of being',[15] that is our original state of being and self knowledge when we walked in the cool of the garden with God as our companion and friend. This theme of restoration is one which flows in and out of eastern theology, often overlaid by a more rigorist and penitential understanding of the 'fall', but like a gentle and strong heartbeat it maintains a rhythm of its own in the comparative short life of the Christian search for God.

Why have I chosen this small interpretive hint? Because it points out that amongst the obvious ways of monasticism viewed in the Church there are other patterns often unnoticed which provide a helpful model for 'thinking outside the box!' In G K Chesterton's 'Father Brown' stories, the little detective priest uses his knowledge both of personal faith and folly to identify the truth in a rational way, looking carefully at the evidence which is almost always right in front of everyone, but goes unperceived because people find it difficult to shift their horizon easily and instead of re-examining the evidence before them, often seek the answer outside of their situation. A question he asks his ex convict friend Flambeau in the story of 'The Sign of the Broken Sword' illustrates this clearly: Father Brown asks 'Where does a wise man hide a pebble?' And Flambeau answers: 'On the beach'. He then asks again: 'Where does a wise man hide a leaf?' Flambeau answers: 'In the forest'.[16] In other words where everything else like them can be found and only the one who knows what they are looking for will see it. The same is true of the monastic experience, where can we find other experiences beyond and different to those we know? The answer is 'in the experience' of the tradition. It is the hints and guesses that often provide the clue. Just because these ways of life are not practised now does not mean that they never did or that in the early stages of a particular form of life, for example Beguines, Friars, a more fluid approach was adopted before Church or political institutionalisation crept up on them. Or, looking at the Eastern patterns of monastic life and some in the West, the understanding that within what at first appears to be a fairly homogeneous structure, some quite original and idiosyncratic life styles emerged. The fact that one can even use the expression 'different types of monk' points out this richness, and though Benedict in his Rule expresses approbation of two types, *Coenobite* and *Hermit* and disapprobation of two others, *Gyrovague* and *Sarabite*, he nevertheless indicates a pattern of alternative behaviours and customs.[17] Just because he castigates the latter two does not mean that there is not another lens to view these in a more constructive light. In fact the wandering fool became a firmly fixed pattern of '*monos*' in the Russian tradition and we can find what appear to be *sarabites* (but only to the outsider) in the desert tradition, the Irish *peregrini* and experience of the 'elder and *staretz*' in the monastic patterns of Orthodox Christianity where individual journeys deviate from the normal pattern.[18] This is quite evident in the early evolution of monasticism which shows the gradual emergence of particular forms rather than any preconceived package. In the First *Sahidic* 'Life of Pachomius', who is credited with the first formation of coenobitic monasticism we get a hint that the ascetic tradition of the hermit evolves into an *ascesis* of more common life which is formed by their spiritual discipline:

> The idea of koinonia, distinctive for the Pachomian monastic movements, denotes something held in common. The idea of mutual support in the religious life implicit in this term-and in the 'Life's' use of the text from Acts—suggests that one of the s chief strengths of the movement founded by Pachomius was the way in which its monks, among them many recent converts like Pachomius himself, could sustain themselves in this ascesis.[19]

How can this aid us in our ecological struggles? It is quite clearly that stress on *koinonia*, community of some kind in which the greater good of all is linked to a more creative understanding of a common *ascesis* of life, a balanced and more simpler way of life more attuned to whole of creation and living beings on our planet! This is of course at the heart of eastern theology, the great theological understanding of the Kingdom present now but not yet, the Kingdom understood and known to us as communion and our vocational role on the planet as people who with, in, and through Christ are drawn together with all creation in the dance of life. The *ascesis* of communion is something the monastic life can offer the world as a model, John Chrsyssavgis put it well when he at states:

'The ascetic way is a way of liberation. And the ascetic is the person who is free, uncontrolled by attitudes that abuse the world; uncompelled by ways that use the world; characterised by self-control, by self-restraint and by the ability to say "no" or "enough". Asceticism, then, aims at refinement, not detachment or destruction. Its goal is moderation, not repression. Its content is positive, not negative; it looks to service, not selfishness; to reconciliation, not renunciation or escape.'[20] If the monastic experience teaches anything it is about continual 'trying', for as with all things human, this way of life is not a perfect way, it is as one English Benedictine Abbot, Dom Gregory Freeman of Douai Abbey once stated, 'a failure, but a worthwhile failure'. What I think he meant was the obvious and typical human desire to exalt particular ways from being 'a way' to 'the way' of following God is the problem, the only antidote is that constant search for truth which is really at the heart of the monastic spiritual quest. The problem is, as we can see with our ecology, that the balance of the world has been upset. We humans need to heed the call to repentance, reverse our perspectives and attitudes, radically change our life styles, what Chryssvagis calls 'salvation through self-denial, the denial of selfishness or self-centredness'.[21]

IV. Christ as Our Word

There is of course an element of conjecture about some of these earlier historical forms of *monos*, records do not often tell us very much, but it would be invidious not to explore the hinterlands and little known areas of our christian spiritual life and look, not for things to resurrect, but for examples to use as a guide and pattern for the 'today' of any age. However there needs to be a cautionary voice in our ears. Not everything from the countryside of history is good, some failures happened precisely because they were unreasonable, ill advised, contrary to the Gospel and other negative things. Others existed for a particular reason at a particular time (the Monk Knights of the Middle Ages) and when their relevance was questionable, either mutated into another form of life of disappeared. But there are also wider issues to consider, especially the areas that concern human experience and growth as well as theological reflection at many levels, for Christianity always turns to theology in order to examine its mission and roots in terms of a God who is triune and who interplays with all life on our small blue planet, not least because of a connectedness through creation, but also through the Incarnation of Jesus Christ and the continuing work of the Holy Spirit. Christian monasticism cannot make any claim for itself outside of the Gospel—here we are reminded very forcibly of that Word, both the Word made flesh, but also the words of scripture that form so much of the spiritual diet of the *monos*. I have already noted one of the prime texts in monastic literature concerning Anthony of Egypt, held up as the great example of monastic existence, hearing the words of the Gospel about the rich young man (Mt 19: 16–19) he heard the Word, the call of Christ as a living encounter and remained referential to that all his life. The innate problem is a human tendency to go beyond that simple call and encounter, to become driven and often far too self referential and inward looking. As the wisdom of the eastern monks and nuns puts it, once we deviate too far from this encounter with Christ and the Spirit, we enter a different world where we can easily get side tracked into thinking that somehow the monastic expression is a more 'spiritual' way or needs to be more extra-ordinary or is the real reflection of the Kingdom of God unlike ordinary human life. There is an antidote, returning to the source of Christ as the true monastic experience constantly does, Pope Francis expresses this well in his Encyclical, *Laudato Si*:

Jesus lived in full harmony with creation, and others were amazed: 'What sort of man is this, that even the winds and the sea obey him?' (Mt 8: 27). His appearance was not that of an ascetic set apart from the world, nor of an enemy to the pleasant things of life. Of himself he said: 'The Son of Man came eating and drinking and they say, "Look, a glutton and a drunkard!"' (*Mt* 11: 19). He was far removed from philosophies which despised the body, matter and the things of the world. Such unhealthy dualisms, nonetheless, left a mark on certain Christian thinkers in the course of history and disfigured the Gospel.

Jesus never said that the monastic or married life was 'the way' instead it is He who is 'the way, the truth and the life' and it is through the Holy Spirit that we discern and discover His presence in our world. Luke's account of the Journey to Emmaus is a good example of this discernment and discipleship, for those disciples the ministry and mission of Jesus was a failure, they were leaving Jerusalem and presumably trying to seek their old familiar patterns of life before Jesus had called them to follow him, tried and tested is best! But it doesn't work, the weight of the living scripture opens their hearts and minds to a new understanding of how the Lord works through life (Moses and the Prophets) then in the breaking of the bread they discover that Jesus wasn't simply with them on their journey but is now always with them, calling them to follow him in proclamation and living out of the Kingdom.

Bonnie Thurston is an anchoress from West Virginia and in a contribution to a collection of essays about the Cistercian Monastic Vision for the twenty-first century, put forward what I think is an excellent synopsis of re-calibrated vows, the four monastic 'Benedictine' vows of poverty, chastity, obedience and stability (*conversio morum* implicitly includes both poverty in some form and chastity). In one sense the lived expression of these vows indicates a charism that can be expressed as gift. Thurston places before us three main 'gifts' of monasticism; 'hospitality, openheartedness, and detachment. These three attitudes are intrinsically inter-related and are the fruit of monastic profession when it is well lived by the individual and the community'.[22] But do these gifts come from and depend entirely on a manner of life based entirely on those four vow areas? I would suggest not always, especially if we look at the fruits of history. To fulfil its own gift, a fruit cannot just remain on a bush or tree or flower, if it isn't harvested in some way then it rots, for it needs to be used, eaten as nourishment for life or seeded to propagate more fruit in its turn and this may also mean changes and responses to differing circumstances, for nothing remains the same for ever except God, and we are people of the twenty-first century with significant expansion in knowledge and understanding of just what it means to be human. The widening of our monastic DNA chart takes us out of categories such as a purely platonic world view with an unfortunate tendency to dualism and the advance of the human sciences warns us about detaching human and divine, something that the incarnation and resurrection of Christ unites. I would suggest that any new lay monastic experience, in whatever manner we find ourselves called to pursue it, can be discerned in these three gifts, ways in which we can seek God by fulfilment of the great commandment of loving God, and our neighbour as we have to love ourselves.

V. The Gift of Silence

In an age of mass communication, the advent of the computer and mobile phone has transformed the earth. We accept without demur instant news, immediate communication. The good side of this is immediately apparent, but more and more questions are being asked about the down side of these technologies, not only in monitoring them so they cannot be misused, but in their contribution to enhancing society and world. It is too early to look at the health issues surrounding these technological developments, but the evidence of shifting paradigms of behaviour are more apparent. The spiritual insights of the past are important for the formation of our responses to new innovations. One such is the crowding out of a true understanding of silence, so essential for monastic life. Paolo Torresan in an article on 'Silence in the Bible' in the Jewish Biblical Quarterly, perceives that the silence of God is understood in three ways, the psychological, historical and the metaphysical.[23] In many of these contexts silence is denoted as absence, even death, but there is also the retrieval of silence as joy and communication. Monastic writers have long seen the famous passage about earthquake, wind and fire, then the silence of the the voice of God found in the First Book of Kings, as a metaphor for the love and essence of what silence in monastic life is all about, this is the silence of joy. In this biblical image, God, in His love for Zion, shouts out with joy and then falls into an enchanted silence: 'He will shout with joy for you, He will jump for you in jubilation, He will be silent in His love' (Zeph. 3: 17). The prophet Elijah had a profound experience of Divine silence at Mount Horeb. Elijah lived at a time when religious syncretism profoundly undermined the faith of Israel because of a policy favored by the Royal Court. Elijah worked to re-awaken the monotheistic faith of the people. At the famous contest on Mount Carmel, the prophets of Baal were slaughtered in the presence of the King. Jezebel threatened Elijah's life and he fled to

Mount Horeb (I Kg. 19: 5–7), the same mountain on which, centuries before, God had appeared to Moses. Here, he had a remarkable theophany. A strong wind passed, then a fire, then an earthquake, but in none of these phenomena was God present. At the end of these manifestations, Elijah perceived *kol demamah dakah*—'a still small voice,' or, in an alternative translation, 'a thin sound of silence.' The prophet, covering his face as a gesture of respect, understood that God had finally shown Himself (I Kings 19: 11–13). This thin veil of joy, the interstices between the Divine realm and ourselves is what the true silence of Christian religious tradition, exemplified in the monastic approach is all about, a communication of the depths! The problem is of course, the issue of modern technology. In the past twenty years mobile phones have totally transformed a cultural manner of engagement, speaking is no longer done face to face, instead the apparently one-sided voice of the mobile phone is heard at any time, in any place, speaking with another often for a long time and alas not always on matters of great moment. It can be very intrusive to others, breaking in on social and private space at inopportune moments. Pope Francis highlights this:

> True wisdom, as the fruit of self-examination, dialogue and generous encounter between persons, is not acquired by a mere accumulation of data which eventually leads to overload and confusion, a sort of mental pollution. Real relationships with others, with all the challenges they entail, now tend to be replaced by a type of internet communication which enables us to choose or eliminate relationships at whim, thus giving rise to a new type of contrived emotion which has more to do with devices and displays than with other people and with nature. Today's media do enable us to communicate and to share our knowledge and affections. Yet at times they also shield us from direct contact with the pain, the fears and the joys of others and the complexity of their personal experiences.[24]

What he writes is simply a consolidation of ancient wisdom from the spiritual tradition. But there is also the problem with our media conscious age, demanding instant access to news does not always mean accurate or intelligent reportage. Some of our morning news programmes fill the gaps with some incredibly naive or simplistic stories, the banter between the different anchors, often banal in the extreme, yet admittedly still popular because people seem to love chat, seem to want any silence crowded out with white noise. This is not healthy, we lose both the instinct to discern more by tuning in to the deeper noise of the world blotted out by our soundings, but also the capacity to listen to ourselves as well. The monks knew this, Abba Peomen expresses it in another manner:

> A man may seem to be silent, but if his heart is condemning others he is babbling ceaselessly. But there may be another who talks from morning till night and yet he is truly silent; that is, he says nothing that is not profitable.[25]

This saying is part of a very balanced approach to an integrative personality in tune with self and world, but one not easy to achieve, it is not only part of the *ascesis* of monasticism but also the discipline of life. It is of course the antithesis of the 'contrived emotion' Pope Francis is criticising. The heart in the past, as also the mind and the soul can be a baffling tool of emotions and words. But there is hope, nature has its own silence and sounds, here they form a different symphony from the cacophony of human creation. This silence of nature, a poignant living silence within which sounds speak of essential communication can be discovered through the lived practice of silence, that way of life so loved by the desert monks, for them silence is as an integral part of life: Peomen points out, 'if you are silent, you will have peace wherever you live.'[26] We can see the discernment and teaching of integrated silence in some of these sayings:

'A brother who lived with some other brothers asked Abba Bessarion, "What ought I to do?" The old man said to him, "Keep silence and do not always be comparing yourself with others"'.[27] Abba Bessarion, at the point of death, said, 'The monk ought to be as the Cherubim and the Seraphim: all eye.'[28] In this sense we are brought back to the connectivity of silence, the symbols of the heavenly Kingdom are always part of the liturgy and prayer tradition of the Eastern Church, here in Bessarion's comment we note that though the search for God is partly done alone, in solitude, it does not cut one off from others, it is the deep listening of true communion with the ground of all being and it must be prepared for by that kenosis, emptying of self as Christ did, which then helps us see the greater connection of God with all things. All prayer, even the prayer of silence, has a liturgical underpinning. Why? Fr John Mack explains it well:

> All prayer, to be true prayer, must be Eucharistic. This certainly means prayer must flow out of a thankful heart. Before it becomes intercession, prayer is first a response to grace received.... It is this drive for communication which also links our prayer to that of others. Thankful hearts are driven to declare their thankfulness to others.... Union with others takes place most truly in the context of shared thanksgiving.[29]

This perhaps is one of the most important perspectives that the Christian tradition of the East can share with the world, a shared sense of responsible humility in which the human family look again at themselves and their companions in life, it won't be easy because vested interests and what we might call 'sins' (to use their term) and failings block progress, but we can try and in doing so start to inculcate the theme of a 'thankful heart' because this shared thanksgiving is really another version of the love we all seek!

VI. Hesychasm: Part of the Gift of Silence

The ideal of an 'integrated' life which is part of this eastern insight and vision of the practise of silence and prayer may not necessarily be so in practice, that is of course the vagaries of human nature, but the zeal and the underlying respect for the world as a place to encounter God and as such loved by God keeps on appearing within the tradition of the Eastern Christians. This is important when one is considering just how we can shape and change attitudes, these monastics remind us that we have ever been the same, learning, technology, intellectual advances all avail us little if we have not got the corresponding humility to know that we do not understand everything, least of all our neighbours the animals and even Mother Earth herself!. Abba Gregory the Theologian gives us a pithy saying reminiscent of the Prophet Micah's threefold dictum; 'This is what the Lord asks of you: only this, to act justly, to love tenderly, and to walk humbly with your God' (Micah 6: 8). 'Abba Gregory said, "These three things God requires of all the baptized: right faith in the heart, truth on the tongue, temperance in the body".'[30] It is easy to see how we can transpose Micah's insight with Gregory's comment, the monk places before us the actions necessary to fulfil the command of God, all of which will lead to love. One major spiritual development of these practices is found in the hesychastic tradition, at first it may seem an exotic perspective, but that would be to ignore its potential as a method of altering our horizons in what we will discover is actually a very simple way!

The Hesychastic tradition of Athonite Monastic tradition moves silence into a new sphere, where repetition of a prayer, the 'Jesus Prayer' leads one into the journey of encounter with the uncreated light that is God, the light of the 'Transfiguration', the icon of which shows in image what we are describing in words, this method of prayer centres on the enlightenment of the self, what in Orthodox theology is called 'being like God', in other words the 'deification of human beings'. It is a very ancient method of prayer done in stillness, in quiet times and quiet places and reached its greatest development, but also a time of crisis in the fourteenth century when it was questioned, challenged and eventually reaffirmed as a true method of deep prayer.[31] Archbishop Chrysostomos of Romania in a lecture given on Hesychasm, explored the theology of the Hesychastic movement, one of his insights is very helpful for placing this spiritual tradition within the frame-work of created life, grounding it as it were:

> Father John Romanides points out, in the Eastern Church, there is no ultimate distinction between the contemplative and active Christian life; moreover, 'while in the state of noetic prayer or glory…one attains to such physical resources that one resists the normal effects of the environment.' In short, Hesychasm leads to a life of spiritual realism and active confrontation with the evils of the world, not to contempt of the world and a retreat into mere contemplation.[32]

It is not difficult to trace the holistic spirituality of contemplative silence in the hesychastic prayer tradition of the East. The goal of hesychasm is the achievement of communion with God, called deification (Latin) and *theosis* (Greek). This is not to be understood in any way as humans becoming God, the Eastern tradition is very clear about its meaning, saints such as St. Cyril the Great (c. 376–444) and St. Gregory Palamas (1296–1359) understood that union with God or '*theosis*' means being partakers or sharers of the divine nature (2 Pet 1: 4). It is participating in the essence of God in a very real way, without humans becoming identical with God, for this can only be achieved through the gift and grace of the Holy Spirit. We begin by the practice of this style of prayer as part of a disciplined life which includes practices that bring body, mind and spirit into a harmony, like the rhythm of the Jesus Prayer where a short sentence, but one of great spiritual power; 'Lord Jesus Christ, Son of the living God: Have mercy on me a sinner' is accompanied by breathing in on the first half of the phrase and out on the second, this repeated continually. It is also a disciplined prayer in the sense that it cannot be done without some preparation and also with the monitoring presence of someone well versed in the spiritual craft: but in time as the elders and saints explain, we then discover that words do not matter, we pass beyond them so that the thin veil of silence becomes the dialogue itself:

> For that reason, hesychasts 'speak about the experience of the Holy Spirit in a very realistic manner, but at the same time they stress that God's nature remains totally "inaccessible" It can be inferred that the spiritual fulfilment of hesychasm in the East is to experience and receive the shared energy of God that brings shalom—a feeling of contentment, completeness, wholeness, well-being and harmony—to human beings as well as the whole creation. Being deified by contemplation, hesychasts receive the wisdom of God here on earth. St. Climacus said, "Let us hear and wonder at

the wisdom of God found in earthen vessels." Nevertheless, all the contemplative practices in the East and the West, in ancient and in current times, arrive at one conclusion of faith: that in contemplation "God completes the work" through "God's grace".'[33]

Notes

1. Pope Francis. Encyclical *Laudato Si*. Vatican, Rome, June 19th 2015. p 126.
2. Schaff, Philip (ed.). 2004. *Nicene and Post Nicene Fathers, Chrysostom*. First Series, Fourth Printing, Volume 13. T&T Clark, London. p 447.
3. Lubheid, C. trans. 1985. *John Cassian Conferences*. Classics of Western Spirituality. Paulist Press, Mahwah, NJ. pp 62, 64.
4. See the cartoon series in, Brother Choleric (Dom Hubert Van Zeller OSB). 1960. *Last Cracks in Legendary Cloisters*. Sheed and Ward, London.
5. Van Zeller, p 443.
6. Luibheid 1985, p 91.
7. Ibid, p 88.
8. Russell, N. trans. 1981. *The Lives of the Desert Fathers*. Mowbray, Oxford. p 45.
9. Ward, B. 1984. *The Sayings of the Desert Fathers*. Cistercian Publications, Kalamazoo, MI, p 21.
10. Maynard, J. LO. 2010.*The Atlas Martyrs*.CTS, London. p 41.
11. Thompson, Francis. 1917. *The Hound of Heaven* in Nicholson & Lee (eds.). *The Oxford Book of English Mystical Verse*. The Clarendon Press, Oxford n. 239.
12. Dunn, M.2003 *The Emergence of Monasticism From the Desert Fathers to the Early Middle Ages*. Blackwell Publishing, Oxford. See Chapter 1.*The Emergence of Christian Eremitism*. pp 4–24.
13. The Benedictine community of Edgeware. This is nothing new, Dom Denys Rutledge in his famous book on the vision and vocation of monastic life, examines a number of shifts and changes but constantly affirms that the Benedictine coenobitic way is the 'normal' way of monastic life. Rutledge, D. 1964. *The Complete Monk, Vocation of the Monastic Order*. Routledge and Kegan Paul, London. Particularly chapter 8, *The Normal Monastery in Tradition*. pp 151–195.
14. Newman, John Henry. 1989. *Essay on Christian Doctrine*. Ian Kerr's edition. Notre Dame Press, Notre Dame, IN. pp 19–40.
15. Dunn 2003, p 8.
16. Chesterton, G.K. 1992. *Father Brown, Selected Stories*. Wordsworth Classics. Ware, Hertfordshire. pp 134–135.
17. Dunn 2003, pp 12,13.
18. A good example of this can be found in a short sociological analysis of the Coptic monastic understanding of 'roles' of behaviour in their tradition, roles that are derived from an individual's talent, interest or vocation but not position. See Meinardus, Otto F.A. 1992. *Monks and Monasteries of the Egyptian Deserts*. The American University in Cairo Press.

Chapter 14. *The Sociology of Coptic Monasticsm*. pp 190–200. For a short account of the Irish *peregrini* see Dunn 2003, pp 140–149.

19. Dunn 2003, p 27.
20. Chryssavgis, J. 2004. Light Through Darkness: The orthodox Tradition. Darton, Longman, Todd. London. p 117.
21. Chryssavgis 2004, p 117.
22. Thurston, Bonnie. 2006. '*Soli Deo Placere Desiderans*' in Patrick Hart OCSO (ed). A *Monastic Vision for the 21st Century. Where Do We Go From Here?* Cistercian Publications, Kalamazoo, MI. pp 8,9.
23. Torresan, Paolo. *Silence in the Bible*. Jewish Biblical Quarterly 31, no. 3 (2003), pp 153–160.
24. LS 47.
25. Ward 1984. p 171.
26. Ibid, p 178.
27. *Ibid*.
28. Ward 1984, p 42.
29. Mack, J. 1999. *Ascending the Heights: A Layman's Guide to The Ladder of Divine Ascent*. Conciliar Press, Ben Lomond, CA. p 161.
30. Ibid, p 44.
31. For a comprehensive overview see Ware, Kallistos. 1992. '*The Hesychasts: Gregory of Sinai, Gregory Palamas, Nicholas Cabasilias*' in Cheslyn Jones, Geoffrey Wainwright, Edward Yarnold SJ (eds). *The Study of Spirituality*. SPCK ,London. pp 242–258.
32. Chrysostomos, Archbishop. 2001. *Orthodox and Roman Catholic Relations from the Fourth Crusade to the Hesychastic Controversy*. Center for Traditionalist Orthodox Studies, Etna, CA. pp 203–204.
    Romanian version: Arhiepiscop Chrysostomos, *Relaţiile Dintre Ortodocşi şi Romano-Catolici de la Cruciada a ,IV-a la Controversa Isihastă*, trans. Raluca Popescu and Mihaela Precup (Bucharest, Romania: Editura Vremea 2001), pp 195–231.
33. The Asbury Journal 70/1:76–94
    © 2015 Asbury Theological Seminary
    DOI: 10.7252/Journal.01.2015S.05
    Moe Moe Nyunt
    Hesychasm Encounters Lectio Divina: An Intercultural Analysis of Eastern and Western Christian Contemplative Practices, pp 83/4.

CONCLUSION

One of the defining perspectives about our human relationship to this wonderful world and all its creatures, found on our journey into Eastern Christianity can be summed up by John of Damascus; 'All, indeed are for the seasonable use of man: but of them some are for food...others for service...and others for enjoyment'.[1] This viewpoint is one about the right sense of balance and proportion that we need to develop towards the earth and its products, including ourselves, and also points out the limitations of our position in it! Though we are integral to the life of our planet, we are not the exclusive owners and users, we are in fact co-dependent with so many other species and they with us. Neither in humility can we say that historically we have been here *ab inititio*, the traditional stories of creation, including those of Genesis firmly put us in position as newcomers whose role as intelligent beings is to care for this little planet.

In our exploration of the various perspectives that can help us look again both with renewed hope and a new sense of our place as God's caretakers, the images and themes from the East help us realize that deep within Christianity there is an all encompassing love that comes from the theology of the Incarnation, a call to a very definite vocation, for as God in wisdom stepped into human and earthly life through Jesus, we are asked to share in that task of conversion and renewal through the gifts of the Holy Spirit. Transformed and renewed through the death and resurrection of the cosmic Lord Jesus, the Kingdom of God now sings out from every page of history on this little planet. Our task is to be people of thanksgiving, exemplified by those deep and rich theologies found in the Eucharist and the Liturgy itself, but also to be people of reconciliation and healing, whose care is for all, especially the least of all creatures. As the human being is the icon of Christ so the world and all its creatures is the icon of heaven, hinting at what is finally to be, a state of life where pain and suffering will be no more. The future lies in our hands, it is hoped that such spiritual teachings from this ancient tradition of Christianity may be part of the blessing we share.

In the end perhaps the greatest perception we may have gained is summed up by Gregory of Nyssa's phrase, 'Love is the mark or feature of our nature from God'.[2] If we truly seek to live out that feature of our vocations on this small planet which we call home, then just perhaps, the world and all its creatures might have a brighter future than we deserve.

Notes

1. Schaff, Nicene and Post-Nicene Fathers Second Series Volume 9, p 28.
2. Ibid, Volume 5, p 391.

INDEX

A

B

C

H

I

J

K

L

M

N

P

R

S

T

V

W

Washington College

Studies in Religion, Politics, and Culture

Joseph Prud'homme
General Editor

Washington College Studies in Religion, Politics, and Culture explores the role of religious belief in political and cultural life, both in the United States and across the world. Special emphasis is placed on theoretical, historical, and legal assessments of the relationship of religion to public life, the meaning and practice of citizenship, and cultural development. Empirical work is also welcome.

For additional information about this series or for the submission of manuscripts, please contact:

Dr. Joseph Prud'homme
Director, Institute for Religion, Politics, and Culture
Washington College
300 Washington Avenue
Chestertown, MD 21620
E-mail: jprudhomme2@washcoll.edu

To order other books in this series, please contact our Customer Service Department:

(800) 770-LANG (within the U.S.)
(212) 647-7706 (outside the U.S.)
(212) 647-7707 FAX

Or browse online by series at www.peterlang.com

www.ingramcontent.com/pod-product-compliance
Ingram Content Group UK Ltd.
Pitfield, Milton Keynes, MK11 3LW, UK
UKHW020207290926
15470UKWH00001B/56

9 781433 153860